高等学校“十三五”规划教材

化学类专业实验

双语版

宋广亮　庄玲华　主编

化学工业出版社

·北京·

内 容 简 介

《化学类专业实验（双语版）》为中英双语版，在内容编排上加入了详尽的实验安全教育方面的内容，结合学科发展融入大量的现代仪器分析技术、新的制备分离技术，并增大了综合实验的比例。全书共两章，第一章为化学实验室基础知识（实验要求和实验室安全）；第二章为实验部分，共编排实验24个，以综合实验为主，实验内容涵盖新材料、有机合成、高分子合成、无机化合物制备、天然产物分离与分析、绿色化学等相关主题，将化合物的制备与分离、产物结构表征和性能测试有机结合并系统化，锻炼提升学生设计综合实验研究的能力。

本书可作为高等院校化学、化工类及相关专业的基础化学实验课教材，也可供从事化学、应用化学等研究的科技人员参考。

图书在版编目（CIP）数据

化学类专业实验：汉英对照/宋广亮，庄玲华主编．—北京：化学工业出版社，2021.3（2025.8重印）
高等学校“十三五”规划教材
ISBN 978-7-122-38279-5

Ⅰ.①化… Ⅱ.①宋…②庄… Ⅲ.①化学实验-高等学校-教材-汉、英 Ⅳ.①O6-3

中国版本图书馆CIP数据核字（2020）第265191号

责任编辑：李 琰 宋林青　　装帧设计：史利平
责任校对：李雨晴

出版发行：化学工业出版社（北京市东城区青年湖南街13号 邮政编码100011）
印　　装：北京科印技术咨询服务有限公司数码印刷分部
787mm×1092mm 1/16 印张10½ 字数254千字 2025年8月北京第1版第3次印刷

购书咨询：010-64518888　　售后服务：010-64518899
网　　址：http://www.cip.com.cn
凡购买本书，如有缺损质量问题，本社销售中心负责调换。

定　价：32.00元

前 言

化学类专业实验是为化学、应用化学等化学类专业学生开设的实验课程，是培养和提升学生专业素质和科研能力的重要实践教学环节。在学生完成基础化学实验和仪器分析实验等课程之后开设的专业实验应具有设计性、综合性、拓展性和先进性，可以启发学生创新意识，训练其综合实践能力。因此，随着时代进步和技术发展，化学类专业实验内容的设置需要与时俱进，及时更新内容以适应新时代背景下高等教育和化学类专业的发展。另一方面，双语教学是高等教育改革的重要措施之一，随着我国高等教育逐步国际化，对化学实验类的双语教材的需求越来越多。因此我们编写了这本中英文双语教材。

本教材的特色与创新点主要有以下几点：(1) 在内容设计上，遵循综合应用性、时代性和知识性的原则，结合学科发展融入较多的现代仪器分析技术、新的制备分离技术，让学生既能学习现代仪器的基本原理和适用范围，也能拓展知识面；(2) 加大了综合实验的比例，体现专业科研新方向，进一步提高学生应用专业知识的综合能力，综合实验项目根据教学学时拆分为连续的小实验，提高教学的可操作性；(3) 每个实验项目均加入实验 HSE（Health 健康、Safety 安全、Environment 环境），强化学生的健康、安全、环保意识，建立预防管理机制，以便最大限度地减少实验事故、环境污染和安全事故；(4) 目前化学实验类的双语教材很少，尤其是专业实验的双语教材，这本中英文双语教材符合国际化学生培育理念，特色鲜明，实用性强。

南京工业大学宋广亮（第一部分，实验 9，22～24，Experiment 9，22～24）、庄玲华（前言，实验 19～21，Experiment 19～21）、陈云（实验 1，3，16，Experiment 1，3，16）、裴文博（实验 2，Experiment 2）、徐璇（实验 5，13～15，Experiment 5，13，15）、王建强（实验 4，Experiment 4）、唐永明（实验 7，8，Experiment 7，8）、王强（实验 12，Experiment 12）、解沛忠（实验 11，Experiment 11）、何广科（实验 6，Experiment 6）、李董艳（Experiment 14）、周翼（实验 17，18，Experiment 17，18）、张显波（实验 10，Experiment 10）、张玲玉（前言、附录）、熊丹（前言、附录）等老师参加了本书的编写工作。宋广亮和庄玲华对全书进行了统稿和校稿，张玲玉和熊丹对本书的全部文字内容进行了校对。

由于编者水平有限，书中难免有疏漏和不妥之处，敬请读者批评指正。

编者

2020 年 6 月

目 录

第一章 化学实验室基础知识

化学类专业实验是本科专业教学中的一个重要实践性环节，通过化学类专业实验课程的训练，使学生初步了解和掌握相关的专业实验必备操作技能、常用仪器设备的操作方法和科学研究方法，使学生的实验操作技能和解决实际问题的能力有较大程度的提高，为学生的进一步学习和工作打下坚实的基础。专业实验显然不同于前面的基础实验，实验目的不仅仅是验证某个已知原理、观察某个明显现象或探索某个普遍规律，也是为具有明确的工业或理论背景的化学问题提供有效的解决方案，因而具有综合性、研究性和开放性。通过专业实验的训练，有利于学生系统地理解科学原理，初步具备实验方案设计、规划路线选择、仪器设备选用、实验数据处理、实验报告撰写等方面的能力。学生应认真做好专业实验，勤于动手、勇于钻研、敢于实践，在实验过程中练好扎实的实验基本功，逐步培养独立从事科学研究工作的能力。

一、实验要求和一般流程

根据各个实验的目的和任务组装实验装置，并在规定时间内独立或共同完成实验操作、数据处理、撰写实验报告。整个流程可分为实验预习、实验操作、实验记录、数据处理和实验报告五个部分。

（1）实验预习。为了保证实验的顺利进行，以达到预期目的，要求学生必须做到充分预习，重点预习内容集中在以下几点：根据实验内容和相关资料，明确实验的目的及原理；认真阅读实验讲义，必要时查阅相关文献，掌握实验项目的要求、内容以及所依据的实验原理；了解所需试剂及其相关参数，如物理性质、化学性质和安全性等；掌握实验所用的相关仪器及设备的使用方法，清楚实验流程、操作控制点和关键数据记录。

（2）实验操作。在操作过程中善于思考、勤于动手、敏锐观察、细心操作、深入分析和钻研，要特别注意实验中的异常现象，严格按照实验步骤和仪器操作规程进行，有异常现象和特殊情况及时汇报指导教师，不得私自改动操作步骤。

（3）实验记录。学生必须准备一个实验记录本，及时且如实地记录实验现象和数据，以便对实验现象做出正确的分析和解释，要养成随做实验随记录的良好习惯，切不可等实验结束后凭回忆补写实验记录，更不允许编造实验数据，经指导教师检查并签名后的实验及其原始数据记录才有效。如实是指实验原始记录的真实性，就是怎么做就怎么写，是什么现象就记什么现象。多步骤复杂实验的操作步骤、现象很难和实验讲义完全吻合，稍有差异也是正常的。及时是指实验原始记录的时间性，在多步骤操作中，每做一步实验都需要在操作结束后及时记下来。因为时间原因，也可以在实验前先简单写好操作步骤，实际操作时再记上具体试剂的用量和时间，用笔划掉和预习记录不一样的地方，并在旁边写上正确数据。实验原始记录是分析实验成败的关键，也是科研诚信的基本要求，还关系到我们做人做事的基本原则。

（4）数据处理。包括对实验数据的准确记录以及对测试数据进行分析计算，特别是有效数字的记录和异常数据的处理。数据处理要做到有理有据，特别是要写出数据处理的全过程而不是只写上最后结果。该过程不仅可以提高学生对化学基础知识的掌握程度，还能够提升学生分析问题、归纳问题、解决问题的能力，有利于培养学生科学严谨的工作作风。

（5）实验报告。学生在实验结束后要撰写实验报告，根据实验操作和实验记录，对实验现象作出解释，并写出相关实验原理，或根据实验数据进行处理和计算，注意分析、讨论实验过程中出现的问题，并作出结论。一个完整的实验报告一般应包括以下主要内容。

① 实验名称　实验名称应当正确地表示所做实验的基本意图，让阅读报告的人一目了然；

② 实验目的　实验目的是对实验意图的进一步说明，即简述该实验在科研或生产中的意义与作用；

③ 实验原理　实验原理是具有普遍意义的基本规律，是实验方法的理论依据或实验设计的指导思想，只有明确实验原理，才能真正掌握实验的关键及操作要点；

④ 实验试剂与仪器　实验所需的主要仪器、设备、工具、试剂，包括它们的基本性质、参数等；

⑤ 仪器装置图　实验所用仪器的连接次序和方式，包括实验仪器操作规范及操作技巧等；

⑥ 操作要点　即实验操作顺序，可以用简图、表格、反应式等表示，不必千篇一律，可以画出实验流程图或实验装置图，再配以相应的文字说明，这样能使实验报告简明扼要、清楚明白；

⑦ 实验结果与讨论　实验结果包括实验现象的描述、实验数据和处理结果等，是实验报告的重点内容。实验数据主要指实验过程中直接测量到的读数，处理结果是指用实验数据经过一定的运算所得到的结果。对于实验结果的表述，可以用文字表述，可根据实验目的将原始资料系统化、条理化，用准确的专业术语客观地描述实验现象和结果，要有时间顺序以及各项指标在时间上的关系；也可以用表格或坐标图的方式，使实验结果突出、清晰，便于相互比较，尤其适用于分组较多且各组观察指标一致的实验，使组间异同一目了然。每一图、表应有标题和计量单位，应说明一定的问题。还可以用计算机软件制作的或实验仪器记录仪显示的反映指标变化的趋势图。在实验报告中，可根据实际条件任选一种或几种方法，以得到最佳效果。实验讨论就是根据实验条件等因素分析产生这一结果的原因，对实验结果的可靠性和合理性进行评价，并解释观察到的实验现象；也包括针对实验中遇到的疑难问题，提出自己的见解和对实验方法、技术路线等提出改进意见。

二、实验室安全基础

1. 实验室注意事项

（1）做好实验预习，思想上要高度重视，对实验安全上可能存在的问题要做到心中有数。进入实验室时必须了解实验室规则及实验安全条例和应急预案，熟悉实验室水电气开关位置，掌握安全用具，如灭火器材、沙箱、洗眼器及急救药箱的放置地点和使用方法，并妥善爱护，不得挪用。

（2）爱护公共财物，不得动用未掌握使用方法以及和本次实验无关的设备、试剂；如有损坏，需及时报告指导教师，并填写损坏单。

(3) 实验开始前应检查仪器是否完整无损，装置是否正确，在征得指导教师同意后，才可进行实验；实验结束后要将仪器、试剂、工具等放回指定位置，并经指导教师和实验室管理人员检查无误后方能离开。

(4) 实验中保持实验台的整齐、清洁，实验台上不放与本次实验无关的东西；不在实验室内吃东西，更不能将实验室的器皿用作非实验之用，实验结束后必须洗净双手。

(5) 实验时要严肃认真，仔细观察现象；保持实验室内安静，不能做与实验无关的事；实验进行时，不得擅自离开岗位，要注意观察反应进行的情况，并观察装置有无漏气和破裂等现象，有特殊情况立即向指导教师汇报。

(6) 当进行有可能发生危险的实验时，要根据实验情况采取必要的安全措施，如穿工作服，戴防护眼镜、防毒面具或橡皮手套等，但不能戴隐形眼镜。

(7) 实验完毕离开实验室时，首先清理好自己的实验台，再由值日生对整个实验室进行整理，最后将水、电、气、门、窗等关好，离开前向指导教师和实验室管理人员汇报，得到允许后方可离开。

(8) 使用易燃、易爆药品时，应远离火源；如发生事故，人身安全第一，不要惊慌失措，应积极想办法抢救，同时报告指导教师。

2. 实验室常见事故预防与处理

虽然在之前的各基础实验课中，学生已经接受过有关实验室的安全教育，但是在专业实验过程中，所使用的化学品可能是易燃、易爆、有毒、有腐蚀性的，常常潜藏着如发生爆炸、着火、中毒、灼伤、割伤、触电等事故的危险性；所涉及的大型仪器价格昂贵，若损坏会造成极大的损失。因此，必须严格执行安全操作规程，加强安全措施，防止事故发生。因此，在进行专业实验之前，有必要再次对学生进行安全教育。

(1) 玻璃仪器割伤的预防与处置

实验前仔细检查所用玻璃仪器是否有破损和裂纹，装置是否安装正确妥当；实验中按规定操作，不强行扳、折玻璃仪器，特别是比较紧的磨口处。保证玻璃仪器的完整，可能有危险的操作时要戴上帆布手套。实验结束后小心拆卸反应装置，仔细清洗干净，分类放到指定位置。

如果不慎发生割伤事故，要及时处理，先将伤口处的玻璃碎片取出；若伤口不大，用蒸馏水洗净伤口，用创可贴包扎，或涂上红药水、撒上止血粉后用医用纱布包扎，必要时送医院处理；若伤口较大或割破了主血管，则应用力按住主血管，防止大出血，及时送医院治疗。

(2) 化学品灼伤的预防与处置

实验前仔细预习，对实验中用到的化学试剂做到心中有数；实验中严禁直接接触化学物质，特别是能引起皮肤灼伤的物质。取用有腐蚀性的化学试剂时，应戴上橡皮手套和防护眼镜。

根据不同的灼伤情况需采取不同的处理方法：被酸或碱灼伤时，应立即用大量水冲洗。酸灼伤用1%碳酸钠溶液冲洗，碱灼伤则用1%硼酸溶液冲洗。最后再用水冲洗。严重者要消毒灼伤面，并涂上软膏，及时送医院就医。被溴灼伤时，应立即用2%硫代硫酸钠溶液洗至伤处呈白色，然后用甘油加以按摩。除金属钠、钾外的任何药品溅入眼内，都要立即用大量水冲洗。冲洗后，如果眼睛未恢复正常，应马上送医院就医。

(3) 化学品中毒的预防与处置

化学物质侵入机体引起伤害的途径主要有吸入、食入和经皮肤吸收。但某物质对机体造成的伤害总是同进入体内的量相联系的，不应该简单地说某种物质有毒有害。不恰当的侵入会引起机体生理功能或正常结构的病理改变，危害健康。我国对空气中有害物质的最高容许浓度有明确的规定，以保证处在该环境中的人不致发生急性和慢性职业性危害而维护人体健康。化学物质的毒性常用引起实验动物某种毒性反应所需的剂量表示，如半数致死量或浓度（LD50 或 LC50）即中毒动物半数死亡的剂量或浓度。根据 LD50，化学物质的急性毒性分为剧毒、高毒、中等毒、低毒、微毒五级。实验中药品不要沾在皮肤上，尤其是极毒的药品。实验完毕后应立即洗手。称量任何药品都应使用工具，不得用手直接接触。使用和处理有毒或腐蚀性物质时，应在通风橱中进行，并戴上防护用品，尽可能避免有机物蒸气扩散在实验室内。对沾染过有毒物质的仪器和用具，实验完毕应立即采取适当方法处理以破坏或消除其毒性。不要在实验室进食、饮水。

化学品中毒处理：一般药品溅到手上，通常使用水和乙醇洗去。实验时若有中毒特征，应到空气新鲜的地方休息，最好平卧，同时汇报实验指导教师。出现其他较严重的症状，如头昏、呕吐、瞳孔放大时应及时送往医院。

（4）着火的预防与处置

不能用烧杯或敞口容器盛装易燃物。加热时，应根据实验要求及易燃物的特点选择热源，注意远离明火。严禁用明火进行易燃液体（如乙醚）的蒸馏或回流操作。尽量防止或减少易燃的气体外逸，倾倒易燃液体时要熄灭火源，且注意室内通风，及时排出室内的有机物蒸气。严禁将与水有猛烈反应的物质倒入水槽中，如金属钠，切忌养成一切东西都往水槽里倒的习惯。注意能在空气中自燃的试剂（如煤油中的钾、钠和水中的白磷）的使用与保存。

实验室如果发生了着火事故，首要应保持沉着镇静，切忌惊慌失措，应及时地采取措施，防止事故的扩大。首先，立即熄灭附近所有火源，切断电源，移开未着火的易燃物。然后，根据易燃物的性质和火势设法扑灭。地面或桌面着火时，如火势不大，可用淋湿的抹布来灭火；反应瓶内有机物着火时，可用石棉布或湿布盖住瓶口，火即熄灭；身上着火时，切勿在实验室内乱跑，应就近卧倒，用石棉布等把着火部位包起来，或在地上滚动以灭火焰。不管用哪一种灭火器都是从火的周围开始向中心扑灭。需要注意水在大多数场合下不能用来扑灭有机物的着火，因为一般有机物都比水轻，泼水后有机物漂浮在水面上燃烧，火随水流动促其蔓延。

（5）爆炸的预防与处置

常压操作加热反应时，切勿在封闭系统内进行。在反应进行时，必须经常检查仪器装置的各部分有无堵塞现象。减压蒸馏时，不得使用机械强度不大的仪器（如锥形瓶、平底烧瓶、薄壁试管等）。必要时，要戴上防护面罩或防护眼镜。使用易燃易爆物（如氢气、乙炔和过氧化物等）或遇水易燃烧爆炸的物质（如钠、钾等）时，应特别小心，严格按操作规程操作。若反应过于猛烈，要根据不同情况采取冷冻和控制加料速度等方式减缓反应。

气瓶泄漏引起的爆炸可通过快速关闭阀门并采用合适的材料和技术手段堵住泄漏处。化学品包装物发生泄漏，应迅速移至安全区域，并更换。少量泄漏用不可燃的吸收物质（如沙子、泥土）包容和收集泄漏物，并放在容器中等待处理。大量泄漏可采用围堤堵截、稀释与覆盖、包容等方法，做好现场处置并立即报告。

（6）触电的预防和处置

认真学习安全用电知识，提高自己防范触电的能力。发生电器设备故障时，不要自行拆卸，请实验指导教师或实验室管理老师处理。不用潮湿的手接触电器。电源裸露部分应有绝缘装置（如电线接头处应裹上绝缘胶布）。所有电器的金属外壳都应保护接地。实验时，应先连接好电路，再接通电源。实验结束时，先切断电源，再拆线路。修理或安装电器时，应先切断电源。不能用试电笔去试高压电，使用高压电源应有专门的防护措施。

如有人触电，应迅速切断电源，然后进行抢救。首先要使触电者迅速脱离电源，越快越好，触电者未脱离前，救护人员不准用手直接触及触电者。触电者脱离电源后，应判断其神志是否清醒，对症紧急处理。发现伤员立即联系校医院救治，或拨打120急救电话。

三、化学废弃物处理

实验室废弃物是指实验过程中产生的三废物质（废气、废液、固体废物）、实验用品残留物和过期试剂或不明试剂，也包括沾染化学物质的生活垃圾。严禁将可能污染环境的化学废液、废渣倒入普通下水道，或将危险化学废弃物（包括受沾染的容器等）随意弃置、填埋。严禁将危险化学废弃物与生物废弃物、放射性废弃物等其他危险废弃物一起混装回收。

化学废液可按照一般无机物废液、一般有机物废液和含卤有机物废液三类，用专用废液桶进行收集和存放。倾倒废液前，须确认倒入后不会与桶内已有物质发生异常反应（如产生有毒挥发性气体、产生剧烈放热现象等）。判断有危险的，可倒入化学试剂空瓶中，贴上标签，注明主要成分。每次倒入废液桶的废液数量和主要成分须做好记录，以便其他人员倾倒时，可对危险性作出明确判断；对化学性质不稳定的废弃物须先做预处理再行收集；倾倒废液时，注意保持安全距离，防止废液溅到身上，并避免将废液倒出桶外；废液桶须随时盖紧，避免内部液体挥发，危害实验人员健康，并放置于阴凉且远离火源和热源的位置，保持良好通风，避免倾倒；废液桶不得盛装太满，须保留1/10左右的空间。严禁将含有剧毒物质的化学废液倒入上述三类废液桶，须倒入化学试剂空瓶中，贴上标签，标明主要成分。

四、实验室应急药箱

为了对实验室意外事故进行紧急处理，实验室配备急救药箱，常备物品清单如下。

药品：碘酒、碘伏（主要用于消毒）、扑热息痛或百服宁等解热镇痛药、牛黄解毒片、黄连上清丸、藿香正气水、黄连素和氟哌酸胶囊、息斯敏和扑尔敏、云南白药、正红花油、风油精、虫咬酊、麝香跌打风湿膏、烫伤膏、0.5％碳酸氢钠、2％稀硼酸、15.1％醋酸、眼药水、板蓝根冲剂。

器具：体温计、剪刀、小刀（手术刀）等常用物品。

其他：消毒纱布、绷带、药用棉花、棉签、创可贴。

五、实验室 HSE 管理

HSE 是 Health（健康）、Safety（安全）、Environment（环境）的英文缩写，HSE 管理体系指的是健康、安全和环境三位一体的管理体系，通过系统化的预防管理机制，消除各种事故、环境和安全隐患，以便最大限度地减少实验事故、环境污染和安全事故的发生。高校实验室引进 HSE 管理体系，是在原有管理方法基础上的改进，体现出以人为本的管理理念，是建设绿色校园的一项重要任务。

实验室安全管理：高校实验室通常储备和使用各种各样的化学试剂、有毒试剂，甚至包括剧毒试剂、易燃和易爆的强氧化剂、强还原剂等，还会放置一些高温高压设备如马弗炉、烘箱和各种气体钢瓶。这些危险化学品和高温高压设备存放不当、违规操作或处理不当乃至疏于管理，就很容易引起燃烧、爆炸、腐蚀、灼伤、中毒、着火触电等安全事故，甚至造成人员伤亡及财产损失。高校实验室是消防工作的重点部门，也是容易发生事故的主要场所之一，应将安全工作放在首位。实验室需认真做好安全隐患排查工作，实验人员每天下班前需对实验室的水、电、钢瓶、阀门、门窗等进行安全检查，确认无安全隐患后，方可锁门离开。同时，需不定期组织实验室人员学习安全防护的相关理论知识，如有必要可与检测部门共同进行。实验室人员必须掌握相关安全理论知识及防护知识，熟悉应急处理技术及措施。

实验室环境管理：高等院校实验室是教学及科研的实践基地，要认真切实贯彻“安全第一，预防为主”的原则和“谁主管，谁负责”的方针，切实加强实验室环境保护工作。随着高校招生规模的逐步扩大和相关实验科学的不断发展，实验室的数量与日俱增，导致实验室排放的有毒、有害物质迅猛增加，严重超出了环境自身承载能力，应引起各实验室及其主管部门的高度重视，采取相应的保护措施，如通过采用绿色化学实验项目、选用健康、无害的实验材料等，加强环境保护的同时，也要求高校的全体师生树立起环境保护意识，“三废”的处理应进行分类收集，定点存放，由学校有关管理部门委托具有资质的单位进行集中处理处置。严禁将实验过程中产生的会污染环境的废液、废渣随便倒入下水道或随意堆放填埋。

实验室健康管理：在做好实验室的安全和环境保护的同时，还要注意实验人员的自身安全与健康。应注意经常保持实验室通风与除尘，严禁在室内吸烟和饮食等；实验操作时尽可能佩戴相应的防护工具进行；在进实验室工作前，严禁饮酒或饮用其他醇类饮料，以免加快人体对有毒、有害物质的吸收；保持室内空气流动通畅，以免有毒有害气体在实验室内蓄积；避免实验人员因吸入一些挥发性较强的有机试剂或高浓度的蒸气而对人体产生较大危害。有些化学物质在人体内具有蓄积性，可引起慢性中毒，必须特别注意。

Basic Knowledge of Chemical Laboratory

Chemistry experiment is an important practical part in undergraduate teaching. Through the training of chemical experiment course, necessary operation skills, common instruments and equipment, and scientific research methods are preliminarily understood and mastered to lay a solid foundation for students' further study and work. Professional experiments are different from basic experiments. The purpose of the professional experiment is not only to verify a known principle, observe an obvious phenomenon or explore a general law, but also to provide effective solutions for chemical problems with a clear industrial or theoretical background. Therefore, it is comprehensive, research-oriented, and open. Through the professional experiment training, students can understand the scientific principles systematically and have the skills of designing experimental schemes, planing route selection, selecting equipment, processing experimental data, writing an experimental report, and so on. Students should do a good job in professional experiments, be diligent in practice, dare to study, dare to practice, consciously learn the solid basic experimental skills in the process of experiment, and gradually cultivate the ability to engage in scientific research independently.

1. Experimental requirements and general process

According to the purpose and task of each chemical professional experiment, the experimental device should be assembled, and the experimental operation, data processing, and the experimental report should be completed independently or jointly within the specified time. The whole process can be divided into five parts: experiment preview, experiment operation, experiment record, data processing, and experiment report.

1.1 Experiment preview

To ensure the smooth progress of the experiment and achieve the expected purpose, students are required to fully preview. The key preview contents focus on the following points: according to the content of the experiment and related materials, students should make clear the purpose and principle of the experiment, carefully read the experiment handout, consult relevant literature when necessary, and master the requirements, contents and experimental principles of the experimental items. Students should also understand the required reagents and related parameters, such as physical properties, chemical properties, and safety, etc, master the use methods of relevant instruments and equipment used in the experiment, and find out the experimental process, operation control points, and key data records.

1.2 Experimental operation

In the operation process, students should be good at thinking, diligent in practice, keen observation, careful operation, in-depth analysis, and research. Students should pay special attention to the problems of abnormal phenomena in the experiment, and accurately carry out the experiment according to the experimental steps and instrument operation procedures. Any abnormal phenomenon and the special situation shall be reported to the instructor in time, and the operation steps shall not be changed without permission.

1.3 Experimental records

Students must prepare an experimental record notebook to record the experimental phenomena and data timely and truthfully, to make correct analysis and explanation of the experimental phenomena. They should form a good habit of keeping records. It is not allowed to make up experimental data by recollection after the experiment is over. The experiment and its original data record can be effective only after the instructor checks and signs. Truthfulness refers to the authenticity of the original record in the experiment, that is, to write as much as possible and to record whatever phenomenon is. The operation steps and phenomena of the multi-step complex experiment are very difficult to be completely consistent with the experimental handouts, and it is normal for some of them to be slightly different. Timeliness refers to the timeliness of the original record of the experiment. In a multi-step operation, every step should be recorded in time after the operation. Because of the time limit, students can also write down the operation steps first and then record the dosage and time of the specific reagent during the actual operation, stroke out the places that are different from the preview record, and write the correct data next to it. The original record of the experiment is the key to analyze the success or failure of the experiment. It is also the basic requirement of scientific research integrity and is closely related to the principle of behaving.

1.4 Data processing

It includes the accurate recording of experimental data and the analysis and calculation of test data, especially the recording of significant figures and the processing of abnormal data. Data processing should be well-grounded. In particular, students should write the whole process of data processing instead of just the final results. Students' mastery of data processing can not only improve students' mastery of basic chemical knowledge but also improve their ability to analyze, summarize and solve problems, which is conducive to cultivate students' scientific and rigorous work style.

1.5 Experimental report

After the experiment, students should write an experimental report. In this report, students could explain the experimental phenomena according to the experimental operation and experimental records and write down the relevant experimental principles. They could also process and calculate according to the experimental data, pay attention to analyze and discuss the problems in the experiment process, and make a conclusion. A complete experimental report should generally include the following main contents: (1) experiment name: the name of the experiment should correctly indicate the basic intention of the experiment so that the readers of the report can clearly understand it. (2) Experimental purpose: the purpose of the experiment is to further explain the intention of the experiment, that is, to describe the significance and role of the experiment in scientific research or production. (3) Experimental principle: the experimental principle is a basic law with universal significance, which is the theoretical basis of experimental methods or the guiding ideology of experimental design. Only when the principle of the experiment is clear, can we grasp the key points and operation points of the experiment. (4) Experimental reagents and instruments: these include the main instruments, equipment, tools, and reagents required for the experiment, including their basic properties, parameters, etc. (5) Equipment diagram: this includes the connection order and mode of instruments used in the experiment, the requirements of experimental operation specifications, and operation taboos and skills, etc. (6) Key points of operation: that is, the operation sequence of the experiment can be expressed by diagrams, tables, reaction equations, etc. We can draw the flow chart or device diagram of the experiment, and then add the corresponding text description, which not only saves a lot of words but also makes the experiment report concise and clear. (7) Experimental results and discussion: the experimental results include the description of experimental phenomena, experimental data, and processing results, which are the key contents of the experimental report. The experimental data mainly refers to the readings directly measured in the process of the experiment. The processing results refer to the results obtained by a certain calculation with the experimental data. The expression of the experimental results can be expressed in words: according to the purpose of the experiment, the original data can be systematized and organized. The experimental phenomena and results can be objectively described with accurate professional terms, and there should be a time sequence and the relationship between various indicators in time. It can also be used in the form of a table or coordinate diagram. This makes the experimental results prominent and clear, easy to compare with each other, especially suitable for the experiment with more groups and the same observation indexes in each group, so that the similarities and differences between groups can be seen. Each chart and table should have a title and a unit of measurement, and a certain central problem should be explained. It can also be made by some computer software or displayed by an experimental instrument recorder to reflect the trend of changes in indicators. In the experiment report, one or several methods can be selected according to the actual conditions to get the best effect. Experimental discussion is factors to analyze the reasons for this result, to evaluate the reliability and rationality of the experimental results, and to explain the observed experimental phenomena based on the experimental conditions and others. It also includes to put forward their own opinions on the difficult problems encountered in the experiment and to put forward suggestions for improving the experimental methods and technical routes.

2. Fundamentals of laboratory safety

2.1 Laboratory precautions

(1) To prepare for the experiment, we should attach great importance to it and have a clear idea of the possible problems in experimental safety. When entering the laboratory, students must understand the laboratory rules, safety regulations, and emergency plans, be familiar with the position of water and electrical switches in the laboratory, master the place and use methods of safety appliances, such as fire-fighting equipment, sandbox, eye washer, and first-aid kit, and take good care of them, and do not use them.

(2) Take good care of the public property, and do not use equipment and reagents if students are not familiar with the use method or that equipment and reagents have nothing to do with the experiment. If there is any damage, it is necessary to report to the instructor in time and fill in the damage list.

(3) Before the experiment starts, check whether the instrument is intact and whether the device is correct. The experiment can only be carried out with the consent of the instructor. After the experiment, the instrument, drugs, tools, etc. should be put back to the designated position, and students could leave after the instructor and laboratory management personnel check and approve.

(4) During the experiment, please keep the laboratory table and its surroundings neat and clean. Do not put anything irrelevant to the experiment on the test bench. Do not eat in the laboratory, Do not use the laboratory utensils for non-experimental purposes. Wash hands after the experiment.

(5) During the experiment, it is necessary to observe the phenomenon seriously and carefully, keep the laboratory quiet, and do nothing irrelevant to the experiment. Do not leave the post without authorization during the experiment. Pay attention to observe the reaction, check whether the device has air leakage and rupture, and report to the instructor in case of any special situation.

(6) When conducting experiments that may cause danger, necessary safety measures should be taken according to the experimental conditions, such as wearing work clothes, protective glasses, gas masks, or rubber gloves. Contact lenses are not allowed.

(7) When leaving the laboratory after the experiment, first clean up your laboratory, and then arrange the whole laboratory by the students on duty. Finally, close the water, electricity, gas, doors, windows, etc. Report to the instructor and experimental management personnel before leaving, and leave after permission.

(8) When using inflammable and explosive drugs, keep away from the fire source. In case of an accident, personal safety is the first. Do not panic. Students should actively try to rescue, and report to the instructor at the same time.

2.2 Prevention and treatment of common accidents in the laboratory

Although in the previous basic experimental courses, the students have received safety education about the laboratory. But in the process of professional experiments, the chemicals used may be flammable, explosive, toxic, and corrosive, and there are often hidden dangers such as an explosion, fire, poisoning, burns, cuts, electric shock, and other accidents. The large-scale instruments involved are expensive, if damaged that can cause great loss. Therefore, it is necessary to strictly implement the safety operation rules and strengthen the safety measures to prevent accidents. Therefore, before the professional experiment, it is necessary to carry out safety education for students again.

(1) Prevention and treatment of glass instrument cut

Before the experiment, carefully check whether the glass instrument used is damaged and cracked and whether the device is properly installed. During the experiment, operate according to the rules, and do not force to pull or break the glass instrument, especially the relatively tight grinding mouth. Try to ensure the integrity of the glass instrument, and wear canvas gloves when operating in danger. After the experiment, the reaction device should be carefully disassembled, cleaned, and placed in the designated position.

If a cut accident happens accidentally, it is necessary to take out the glass fragments at the wound. If the wound is not large, clean the wound with distilled water, bandage it with a band-aid, or apply red medicinal liquid, sprinkle hemostatic powder, and then wrap it with medical gauze, and send it to the hospital for treatment if necessary. If the wound is large or the main blood vessel is cut, the main blood vessel should be pressed to prevent massive bleeding and the person should be sent to the hospital in time treatment.

(2) Prevention and disposal of chemical burns

Preview carefully before the experiment, and know the chemicals used in the experiment. It is strictly forbidden to directly contact with chemical substances, especially those that can cause skin burns. When taking corrosive chemicals, rubber gloves and protective glasses should be worn.

Different treatment methods should be adopted according to different burn conditions: when burned by acid or alkali, wash with plenty of water immediately. Acid burns were washed with a 1% sodium carbonate solution; alkali burns were washed with a 1% boric acid solution. Finally, rinse with water. In serious cases, the burning surface should be disinfected and applied with ointment, and the person should be sent to the hospital in time. When burned by bromine, wash it with 2% sodium thiosulfate solution immediately until the wound is white, and then massage with glycerin. If you are scalded by the hot glass or ironware, wash the wound with cold tap water for a few minutes or apply ice to the wound until the pain is relieved. For the more serious one, apply safflower oil to the affected area, and then rub some scald ointment. In addition to metal sodium, potassium, any drug splashed into the eye, should be immediately washed with a large amount of water. If eyes do not return to normal after flushing, they should be sent to the hospital immediately.

(3) Prevention and treatment of chemical poisoning

The main ways of injury are inhalation and absorption. However, the damage caused by a substance to the body is always related to the amount of the substance entering the body. It should not be simply said that some substance is toxic and harmful. The improper invasion will cause pathological changes in physiological function or normal structure of the body, which is harmful to health. Proper intake of some substances is not harmful to the human body. In China, there are clear regulations on the maximum allowable concentration of harmful substances in the air, to ensure that people in the environment will not cause acute and chronic occupational hazards and maintain good health. The toxicity of chemical substances is usually expressed by the dose required to cause certain toxic reactions in laboratory animals, such as LD50 or LC50, that is, the dose or concentration of half death of poisoned animals. According to LD50, the acute toxicity of chemical substances can be divided into five grades: acute toxicity, high toxicity, medium toxicity, low toxicity, and slight toxicity. In the experiment, the drugs should not be touched on the skin, especially the extremely toxic drugs. Wash hands immediately after the experiment. Tools should be used to weigh any medicine, and direct hand contact is not allowed. When using and handling toxic or corrosive substances, it should be carried out in the fume hood and wear protective equipment to avoid the diffusion of organic vapor in the laboratory. For instruments and appliances contaminated with toxic substances, appropriate measures should be taken immediately after the experiment to destroy or eliminate their toxicity. Do not eat or drink in the laboratory. Food is easily contaminated with toxic chemicals in the laboratory.

Chemical poisoning treatment: if general drugs splashed on the hand, they could usually be washed away with water and ethanol. If you have poisoning characteristics during the experiment, you should rest in a place with fresh air, preferably lying on your back, and report to the experimental instructor at the same time. If other serious symptoms such as spots, dizziness, vomiting, and mydriasis occur, the person should be sent to the hospital in time.

(4) Fire prevention and disposal

Do not use beakers or open containers to hold flammable materials. When heating, the heat source should be selected according to the experimental requirements and the characteristics of combustible materials, and attention should be paid to keep them away from an open fire. Do not use an open fire for distillation or reflux of flammable liquids (such as ether). Try to prevent or reduce the escape of flammable gas, extinguish the fire source when dumping, and pay attention to indoor ventilation, and timely discharge indoor organic vapor. It is strictly forbidden to pour substances with violent reaction with water into the water tank, such as metal sodium, and avoid pouring everything into the water tank. Pay attention to the use and preservation of some reagents that can spontaneously ignite in the air (such as potassium and sodium in kerosene and white phosphorus in water).

If there is a fire accident in the laboratory, the first thing to do is to keep calm and avoid panic and take measures in time to control the expansion of the accident. First of all, immediately extinguish all nearby fire sources, cut off the power supply, and remove unburned flammable materials. Second, try to put out the fire according to the nature of the in flammables and the fire situation. If there is a fire on the ground or the tabletop, wet rags can be used to extinguish the fire. Asbestos board or wet cloth can be used to cover the mouth of the bottle to extinguish the fire. When your body is on fire, do not run around in the laboratory, but lie down nearby, wrap the fire part with asbestos cloth, or roll on the ground to extinguish the fire. No matter what kind of fire extinguisher is used, the fire should be put out from the surroundings to the center. Note that water cannot be used to extinguish organic fires in most cases. Because ordinary organic matter is lighter than water, after splashing water, the fire does not extinguish but floats on the water surface to burn. When the fire flows with water, it spreads quickly.

(5) Explosion prevention and disposal

Do not operate the heating reaction in a closed system under normal pressure. When the reaction is in progress, it is necessary to check all parts of the instrument and device for blockage. In vacuum distillation, instruments with low mechanical strength (such as a conical flask, flat bottom flask, thin-walled tube, etc.) shall not be used. If necessary, wear a protective mask or protective glasses. When using inflammable and explosive materials (such as hydrogen, acetylene, peroxide, etc.) or substances easy to burn and explode when contacting water (such as sodium, potassium, etc.), special care should be taken and the operation procedures should be strictly followed. If the reaction is too violent, freezing and controlling the feeding speed should be adopted to slow down the reaction according to different conditions.

In case of an explosion caused by leakage of a gas cylinder, the valve can be closed quickly and the leakage shall be blocked by proper materials and technical means. In case of leakage of the chemical package, it shall be moved to a safe area and replaced. A small amount of leakage is contained and collected with nonflammable absorbent materials (such as sand and soil) and placed in containers for disposal. A large amount of leakage can be blocked by a dike, diluted and covered, contained, and other methods, and the on-site disposal and immediate report can be made.

(6) Prevention and disposal of electric shock

Seriously study the knowledge of safe use of electricity and improve your ability to prevent electric shock. In the case of electrical equipment failure, do not disassemble it by yourself. Ask the experiment instructor or the experiment management teacher to deal with it. Do not touch electrical appliances with wet hands. The exposed part of the power supply shall be provided with an insulation device (for example, the wire joint shall be wrapped with insulating tape). The metal enclosure of all electrical appliances shall be grounded. In the experiment, the circuit should be connected first and then the power supply should be connected. At the end of the experiment, cut off the power supply first, and then dismantle the circuit. When repairing or installing electrical appliances, cut off the power supply first. Do not use the test pen to test high-voltage electricity. Special protective measures should be taken when using a high-voltage power supply.

In case of electric shock, the power supply shall be cut off quickly, and then rescue shall be carried out. The first thing to do is to make the person who has an electric shock get rid of the power supply as soon as possible. Before the electric shock is separated, the rescue personnel are not allowed to directly contact the person with electric shock by hand. After the electric shock is disconnected from the power supply, it is necessary to judge whether the person is conscious or not, and deal with the case urgently. If the wounded is found, contact the school hospital for treatment immediately, or dial the emergency call 120.

3. Chemical waste treatment

Laboratory waste refers to the three wastes (waste gas, waste liquid, solid waste) produced in the process of experiment, residues of experimental supplies, expired or unknown reagents, and domestic waste contaminated with chemical substances. It is strictly forbidden to pour the chemical waste liquid and waste residue which may pollute the environment into the common sewer, dispose or bury the hazardous chemical waste (including the contaminated containers) at will. It is strictly forbidden to recycle hazardous chemical wastes together with biological wastes, radioactive wastes, and other hazardous wastes.

The chemical waste liquid can be collected and stored in a special waste tank according to three types: general inorganic waste liquid, general organic waste liquid, and halogenated organic waste liquid. Before dumping the waste liquid, it is necessary to confirm that the waste liquid will not react with the existing substances in the barrel (such as toxic volatile gas, severe heat release, etc.). If it is judged to be dangerous, it can be poured into the empty bottle of chemical reagent and labeled with the main ingredients. The quantity and main components of waste liquid poured into the waste liquid tank must be recorded so that other people can make a clear judgment on the danger when dumping. The wastes with unstable chemical properties must be pretreated before collection. When dumping the waste liquid, be aware to keep a safe distance to prevent the waste liquid from splashing on the body and avoid pouring the waste liquid out of the barrel. The waste liquid tank must be covered tightly at any time to avoid the volatilization of inside the liquid which could endanger the health of laboratory personnel. It should be placed in a cool and cool place far away from fire and heat sources. Keep good ventilation and avoid dumping. The waste liquid tank should not be too full, and about 1/10 of the space should be reserved. It is strictly forbidden to pour the chemical waste liquid containing highly toxic substances into the above-mentioned three kinds of waste liquid tanks. It must be poured into the empty chemical reagent bottle with labels indicating the main components.

4. Laboratory emergency kit

For emergency treatment of laboratory accidents, the laboratory is equipped with a first-aid kit, and the list of standing items is as follows:

Drugs: iodine wine, iodophor (mainly used for disinfection), paracetamol or Bufferin and other antipyretic and analgesic drugs, Niuhuang Jiedu Tablets, Huanglian Shangqing pills, Huoxiang Zhengqi water, Berberine, and Norfloxacin Capsules, Astemizole, and Chlorphenamine, Yunnan Baiyao, Zhenghonghua oil, Fengyoujing, Insect bite tincture, Shexiang Dieda Fengshi ointment, scalding ointment, 0.5% sodium bicarbonate, 2% dilute boric acid, 15.1% acetic acid, eye drops Water and Banlangen Granules.

Equipment: body temperature juice, scissors, knife (scalpel), and other commonly used items.

Others: sterile gauze, bandage, medical cotton, cotton swab, band-aid.

5. Laboratory HSE Management

HSE is the abbreviation of health, safety, and the environment. HSE management system refers to the trinity management system of health, safety, and the environment. A systematic prevention and management mechanism are established to eliminate all kinds of accidents, environment, and safety risks, to minimize the occurrence of experimental accidents, environmental pollution, and safety accidents. The introduction of the HSE management system in the university laboratory is an improvement based on the original management method, reflecting the people-oriented management concept, and is an important task to build a green campus.

Laboratory safety management: university laboratories usually reserve and use various chemical reagents, toxic reagents, even highly toxic reagents, flammable and explosive strong oxidants, strong reducing agents, and other substances, as well as some high temperature and high-pressure equipment such as muffle furnace, oven, and various gas cylinders. If these dangerous chemicals, high-temperature and high-pressure equipment are not properly stored, or there is some illegal operation or improper treatment, or some poor management and other reasons happen, it is easy to cause combustion, explosion, corrosion, burns, poisoning, fire, and electric shock and other safety accidents, and even cause casualties and property losses. The university laboratory is the key department of fire protection work, and also one of the main places prone to accidents. Safety work should be put in the first place from the beginning to the end. The laboratory needs to do a good job in the investigation of potential safety hazards. The laboratory operator should check the water, electricity, cylinders, valves, doors, and windows of the laboratory before going off duty every day. After confirming that there is no potential safety hazard, they can lock the door and leave. At the same time, it is necessary to organize the laboratory personnel to learn the relevant theoretical knowledge of safety protection from time to time. If necessary, it can be carried out jointly with the testing department. Laboratory operator must master relevant safety theory knowledge and protection knowledge, and be familiar with emergency treatment technology and measures.

Laboratory environmental management: the laboratories of colleges and universities are the practice base of teaching and scientific research. We should earnestly implement the principle of "safety first, prevention first" and the policy of "who is in charge, who is responsible", and effectively strengthen the environmental protection of the laboratory. With the gradual expansion of the enrollment scale of colleges and universities and the continuous development of related experimental science, the number of laboratories is increasing day by day, which leads to the rapid increase of toxic and harmful substances discharged by laboratories and that seriously exceeds the environmental carrying capacity of the environment itself. Therefore, the laboratories and their competent departments should attach great importance to it and take corresponding protective measures, such as adopting green experimental projects and selecting appropriate ones. At the same time, all teachers and students in colleges and universities are required to establish an awareness of environmental protection. The treatment of "three wastes" should be classified and collected and stored at fixed points. The relevant management departments of the university should entrust qualified units for centralized treatment. It is strictly forbidden to pour the waste liquid and waste residue which pollutes the environment into the water or pile up for landfill at will.

Laboratory health management: While doing a good job in laboratory safety and environmental protection, we must also pay attention to the safety and health of laboratory operator. Attention should be paid to keep ventilation and dust removal in the laboratory. Smoking and eating are strictly prohibited in the laboratory. Corresponding protective tools should be worn as far as possible during the experimental operation. Alcohol and other alcoholic drinks are strictly prohibited before entering the laboratory to avoid the absorption of toxic and harmful substances by the human body. The indoor airflow should be kept smooth to avoid the accumulation of toxic and harmful gases in the laboratory. The staff should carry out health examinations regularly and treat the disease in time, to prevent the experimental personnel from inhaling some volatile organic reagents or high concentration steam, which could do great harm to human health. Some chemicals are harmful to humans and can cause chronic poisoning, so special attention must be paid.

第二章 实验部分

实验一 溶胶-凝胶法制备纳米二氧化钛

一、实验目的

1. 了解二氧化钛的性质及应用。
2. 掌握溶胶-凝胶法制备二氧化钛的技术。
3. 学习评价纳米二氧化钛光催化活性的一种方式。

二、实验原理

纳米 TiO_2 是一种高功能精细无机材料，具有表面效应、体积效应和量子尺寸效应，是重要的陶瓷、半导体及光催化材料。利用纳米 TiO_2 对有机污染物进行光催化降解，可使有害物质矿化为 CO_2、H_2O 及其他无机小分子物质，因此可用于废水处理、空气净化以及杀菌除臭。对于解决目前日益严重的环境污染问题来说，TiO_2 光催化氧化技术极具研究和实用价值。

1976 年 S. N. Frank 等将半导体材料用于催化光解污染物，取得了突破性的进展。J. H. Carry 报道了 TiO_2 水浊液在近紫外光的照射下可使多氯联苯脱氯，从而开辟了 TiO_2 光催化氧化技术在环保领域的应用。光催化氧化工艺简单、操作条件容易控制、氧化能力强、无二次污染，加之 TiO_2 化学稳定性高、无毒、成本低，故将 TiO_2 作为催化剂的光催化氧化法是一种具有广阔应用前景的废水处理新技术。

溶胶-凝胶法使用含高化学活性组分的化合物作为前驱体，在液相下将这些原料均匀混合，并进行水解、缩合反应，在溶液中形成稳定的透明溶胶体系，溶胶经陈化胶粒间缓慢聚合，形成三维空间网络结构的凝胶，凝胶网络间充满了失去流动性的溶剂。凝胶经过干燥、烧结固化可制备出纳米结构的材料。

溶胶-凝胶法是纳米 TiO_2 的制备方法中很重要的一种方法。它以钛酸丁酯为原料，通过水解和缩聚反应制得溶胶。再进一步缩聚得到凝胶，凝胶经干燥、煅烧得到纳米 TiO_2。其反应方程式为：

$$Ti(OR)_4 + 4H_2O \longrightarrow Ti(OH)_4 + 4ROH$$

$$Ti(OH)_4 \longrightarrow TiO_2 + 2H_2O$$

该工艺原料的纯度较高，整个过程不引入杂质离子，可以通过严格控制工艺条件，制得纯度高、粒径小、粒度分布窄的纳米粉体，且产品质量稳定。缺点是原料成本高，干燥、煅烧时凝胶体积收缩大，易造成纳米 TiO_2 颗粒间的团聚。

本实验以钛酸丁酯为前驱体，无水乙醇为分散剂，冰乙酸为抑制剂，通过溶胶-凝胶工艺制备纳米二氧化钛，并且以甲基橙为模型污染物，进行了纳米二氧化钛光催化活性研究。

三、仪器与试剂

1. 仪器

磁力搅拌器，烧杯，量筒，恒压滴液漏斗，马弗炉，离心机。

2. 试剂

钛酸丁酯（TNB），无水乙醇，冰乙酸。

四、实验步骤

分别量取钛酸丁酯 10mL、无水乙醇 15mL、冰乙酸 6mL，放入烧杯中磁力搅拌均匀，此液体为 A 液。

另量取无水乙醇 4mL、冰乙酸 2mL、二次蒸馏水 4mL 混合均匀，此混合液为 B 液。在剧烈搅拌下，将 B 液缓慢滴入 A 液中。滴加完毕，继续搅拌 3h，静置陈化，待液体失去流动性，在 100℃下烘 12h 得黄色凝胶块，研磨得到无定形纳米 TiO_2。将其放入马弗炉中 500℃下煅烧 2h，研磨得到纳米 TiO_2 晶体。

五、实验 HSE

1. 乙酸：易燃，有强烈气味，具有腐蚀性，对呼吸道有刺激。急性毒性，口服-大鼠 LD50：3530mg/kg，静脉-兔 LD50：1060mg/kg。急救措施：皮肤接触时立即脱去污染的衣着，用大量流动清水冲洗至少 15min；眼睛接触时立即提起眼睑，用大量流动清水或生理盐水彻底冲洗至少 15min；吸入则应迅速脱离现场至空气新鲜处，保持呼吸道通畅，如呼吸困难，给输氧；如呼吸停止，立即进行人工呼吸；如遇食入应立即用水漱口，严重时就医。储存于阴凉、通风的库房。应与强氧化剂、食用化学品等分开存放。保持容器密封，远离火种、热源。

2. 实验中需使用烘箱、马弗炉等高温设备，必须做好防护，防止高温烫伤，全程不得离开实验室。

3. 由于无水乙醇和冰乙酸易挥发，量取样品时请务必在通风橱中进行。

4. 室温较低时冰乙酸易凝固，使用前需要先解冻。

六、思考题

冰乙酸在溶胶-凝胶法中的作用是什么？

Experiment 1 Preparation of Nano-TiO_2 by Sol-Gel Method

1. Objectives

1.1 Understand the properties and applications of TiO_2.

1.2 Master the technology of preparing TiO_2 by sol-gel method.

1.3 Learn a method to evaluate the photocatalytic activity of nano-TiO_2.

2. Principle

Nano-TiO_2 is a high functional inorganic material which always be considered as an important ceramic, semiconductor, and photocatalytic material with surface effect, volume effect, and quantum size effect. The photocatalytic degradation of organic pollutants by nano-TiO_2 can mineralize harmful substances into CO_2, H_2O, and other inorganic small molecules. Therefore, it can be used for wastewater treatment, air purification, sterilization, and deodorization. Nano-TiO_2 has a very great value for solving the increasingly serious environmental pollution problems.

In 1976, S. N. Frank et al. made a breakthrough in the use of semiconductor materials for catalyzing photolysis of pollutants. J. H. Carry reported that TiO_2 suspension can conduct the dechlorination of polyoxy biphenyl under the irradiation of near-ultraviolet light, which creating the possibility of using TiO_2 photocatalytic oxidation technology in environmental protection field. The photocatalytic oxidation process is simple, the operating conditions are easy to control, the oxidation ability is strong, and there is no secondary pollution. In addition, TiO_2 has advantages of high chemical stability, non-toxic, and low cost. Due to all these reasons, we consider the photocatalytic oxidation which use TiO_2 as a catalyst is a new wastewater treatment technology with broad application prospects.

The sol-gel method is a process which uses a compound containing a highly chemically active component as a precursor, uniformly mixing these materials in liquid phase, and using hydrolysis and condensation chemical reactions to form a stable transparent sol system in the solution. The sol slowly polymerizes between the aged colloidal particles to form a gel of a three-dimensional network structure. The gel network is filled with solvents that lose fluidity. After dried and sintered, the gel can prepare a nanostructured material.

The sol-gel method is an important way to prepare nano-TiO_2. It uses a titanium yeast salt as material to prepare a sol by hydrolysis and condensation chemical reactions. Further condensation is carried out to obtain a gel, and the gel is dried and calcined to obtain nano-TiO_2. The reaction equations are as follows:

$$Ti(OR)_4 + 4H_2O \longrightarrow Ti(OH)_4 + 4ROH$$

$$Ti(OH)_4 \longrightarrow TiO_2 + 2H_2O$$

The purity of the material used in this method is high, and no impurity ions are introduced in the whole process. The nano-powder with high purity, small particle size, and narrow particle size distribution can be obtained by strictly controlling the process conditions, and the product quality is stable. The disadvantages of the method are the cost of this method is high and the gel will shrink too much during drying and calcination, which can easily cause agglomeration between the nano-TiO_2 particles.

In this experiment, butyl titanate was used as precursor, anhydrous ethanol was used as dispersant, glacial acetic acid was used as inhibitor, nano-TiO_2 was prepared by sol-gel method, and the photocatalytic activity of nano-TiO_2 was studied by using methyl orange as the pollutant model.

3. Instruments and Reagents

3.1 Instruments

Magnetic stirrer, beakers, measuring cylinder, constant pressure dropping funnel, muffle furnace, centrifuge.

3.2 Reagents

Butyl titanate (TNB), anhydrous ethanol, glacial acetic acid.

4. Procedures

4.1 10mL of tetrabutyl titanate, 15mL of anhydrous ethanol, and 6mL of glacial acetic acid were placed in a beaker and stirred magnetically to form liquid A.

4.2 4mL of anhydrous ethanol, 2mL of glacial acetic acid, and 4mL of redistilled water were mixed well to form B solution. Slowly dripped the solution B into solution A while vigorous stirring. The mixture was stirred for 3 h and after aged, when the liquid lost fluidity, it was baked at 100 ℃ for 12 h to obtain a yellow gel block, and the amorphous nano-TiO_2 was obtained by grinding.

4.3 The material was calcined in a muffle furnace at 500℃ for 2 h, then grind it to obtain nano-TiO_2 crystals.

5. Experimental HSE

5.1 Acetic acid: flammable, strong odor, corrosive, irritating to respiratory tract. Acute toxicity, oral-rat LD50: 3530mg/kg, intravenous-rabbit LD50: 1060mg/kg. Firstaid measures: Immediately remove contaminated clothing and rinse with plenty of flowing water for at least 15 minutes when skin contacts. Lift up the eyelids immediately upon eye contact and rinse thoroughly with plenty of running water or saline for at least 15 minutes. Quickly leave the scene to fresh air to keep the respiratory tract unobstructed when inhalation. Give oxygen when breathing is difficult. If breathing stops, give artificial respiration immediately. If swallowed, gargle immediately with water. Seek medical attention if severe. Store in a cool, ventilated warehouse. It should be stored separately from strong oxidants and edible chemicals. Keep the container sealed and away from fire and heat sources.

5.2 High-temperature equipments such as ovens and muffle furnaces are used in the experiment. It must be properly protected against high-temperature scalds.

5.3 Since absolute ethanol and glacial acetic acid are easy to volatile, be sure to operate in a fume hood when measuring samples.

5.4 Glacial acetic acid is easy to solidify when the room temperature is low, and it needs to be thawed before use.

6. Question

What is the role of glacial acetic acid played in the sol-gel method?

实验二 XRD分析纳米二氧化钛的晶相

一、实验目的

1. 掌握XRD进行物相分析的方法。
2. 熟悉XRD测试的基本操作方法。
3. 了解二氧化钛的三种晶相。

二、实验原理

XRD（X-ray diffraction），即X射线衍射，是一种可快速获得晶态材料的晶相、晶胞等信息的分析技术。由于X射线的波长和晶体内部晶面间距相近，晶体可以作为X射线的空间衍射光栅，即当一束X射线通过晶体时将发生衍射，衍射波叠加的结果是射线的强度在某些方向上加强，在其他方向上减弱。简单来说，通过X射线衍射测试，可以得到特征衍射图谱。这些特征衍射图谱就像化学指纹，通过和数据库里已知数据进行比较分析，可以得出测试样品的相关化学信息。

布拉格方程（$2d\sin\theta = n\lambda$）是衍射分析中最重要的基础公式。其中，d为晶面间距，θ为入射X射线与相应晶面之间的夹角，n为反射级数（$n=1,2,\cdots$），λ为X射线波长。对于X射线衍射，当光程差等于波长的整数倍时，即满足布拉格方程时，将发生衍射，相应晶面的散射线将加强。

从样品的X射线衍射图谱，可以得到以下信息。

✓ 物相：它是X射线衍射应用最多的方面。对材料测得的衍射强度花样与国际粉末衍射标准联合会（JCPDS）收集发行的粉末衍射图谱集卡片相对照，如果和某一标准衍射图谱一致，就可确定材料中存在该物相。

✓ 样品的结晶度：结晶度，即结晶的完整程度。结晶完整的晶体，晶粒较大，衍射谱线强且尖锐，衍射峰的半高宽接近仪器本身的自然宽度；结晶差的晶体，由于晶粒过于细小，晶体中有位错等缺陷，使衍射峰形宽阔而弥散；结晶度越差，衍射能力越弱，衍射峰越宽，直至消失在背景之中。

✓ 晶粒尺寸：Debye-Scherrer德拜-谢乐公式［$D = K\lambda/(\beta\cos\theta)$］，是使用XRD分析晶粒尺寸的著名公式。

式中，K为Scherrer常数；D为晶粒垂直于晶面方向的平均厚度；β为实测样品衍射峰半高宽度；θ为衍射角；λ为X射线波长。由于材料中的晶粒大小并不完全一样，故所得数据实为不同大小晶粒的平均值。

二氧化钛（TiO_2）俗称钛白粉，它是白色固体或粉末状样品，分子量为79.83。广泛用于白色无机颜料，具有无毒、最佳的不透明性、最佳白度和光亮度，被认为是现今世界上性能最好的一种白色颜料。钛白的黏附力强，不易发生化学变化，广泛应用于涂料、塑料、造纸、印刷油墨、化纤、橡胶、化妆品等领域。

二氧化钛有锐钛矿型（Anatase）、金红石型（Rutile）和板钛矿型（Brookite）三种晶型。其中，板钛矿型相对不稳定，也不容易制备。锐钛矿型和金红石型最为常见，这两种晶型均属于四方晶系。

（1）锐钛矿型的标准 XRD 图谱和卡片如下所示。

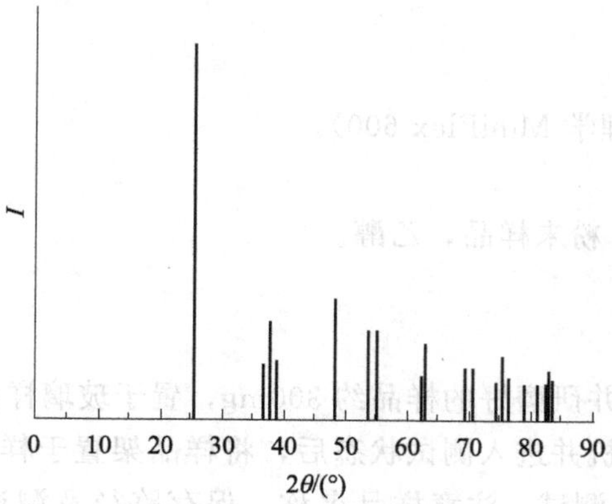

2θ	Int-f	h	k	l	2θ	Int-f	h	k	l	2θ	Int-f	h	k	l
25.325	999	1	0	1	62.165	27	2	1	3	78.754	1	2	0	6
36.989	62	1	0	3	62.750	118	2	0	4	80.860	4	0	0	8
37.841	190	0	0	4	68.842	51	1	1	6	82.236	6	3	0	3
38.601	72	1	1	2	70.346	55	2	2	0	82.759	43	2	2	4
48.074	249	2	0	0	74.155	5	1	0	7	83.232	18	3	1	2
53.952	158	1	0	5	75.129	85	2	1	5					
55.106	155	2	1	1	76.100	23	3	0	1					

Ref：Burdett，J. K et al.，J. Am. Chem. Soc.，109，3639，1987.

（2）金红石型的标准 XRD 图谱和卡片如下所示。

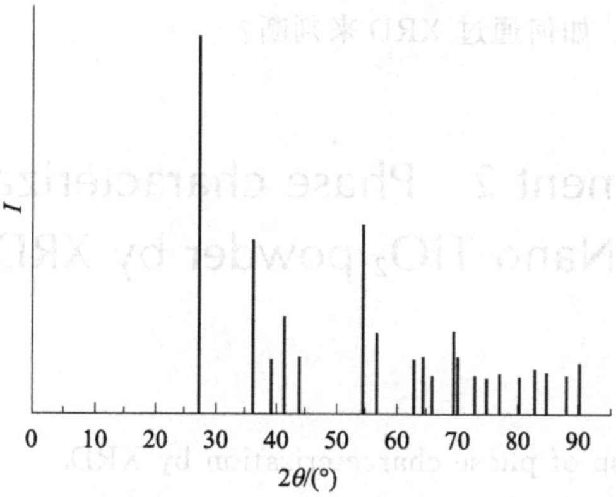

2θ	Int-f	h	k	l	2θ	Int-f	h	k	l	2θ	Int-f	h	k	l
27.459	999	1	1	0	62.799	60	0	0	2	76.597	13	2	0	2
36.104	399	1	0	1	64.105	56	3	1	0	79.894	8	2	1	2
39.223	54	2	0	0	65.565	5	2	2	1	82.414	24	3	2	1
41.269	184	1	1	1	69.064	140	3	0	1	84.333	17	4	0	0
44.081	64	2	1	0	69.851	67	1	1	2	87.568	7	4	1	0
54.364	444	2	1	1	72.483	8	3	1	1	89.628	40	2	2	2
56.676	131	2	2	0	74.470	2	3	2	0					

Ref：Sanchez，E et al.，J. Solid State Chem.，122，309，1996.

本实验分别对商业购买 TiO_2 样品及自制 TiO_2 样品进行 XRD 测试，通过样品衍射花样分析，对比标准卡片，确定各样品的晶相。

三、仪器与试剂

1. 仪器

X 射线衍射仪（日本理学 MiniFlex 600）。

2. 试剂

商业购买及自制 TiO_2 粉末样品，乙醇。

四、实验步骤

1. 装样，取预先干燥并研磨好的样品约 300mg，置于玻璃样品架上，并压平整。

2. 样品测试，仪器开机并进入测试状态后，将样品架置于样品槽内，应用 X 射线粉末衍射仪，对样品进行 XRD 测试。注意样品名称、保存路径及测试条件（如测试角度范围、步长、扫速等参数）的设置和记录。

3. 回收测试样品并关机。

五、实验 HSE

注意 X 射线辐射。

六、思考题

1. TiO_2 有哪些用途？

2. 从样品的 X 射线粉末衍射图能得到材料的哪些信息？

3. TiO_2 有哪些晶相，如何通过 XRD 来判断？

Experiment 2 Phase characterization of Nano-TiO_2 powder by XRD

1. Objectives

1.1 Get a good grasp of phase characterization by XRD.

1.2 Be familiar with the instrument operation for the XRD test.

1.3 Know the three phases of TiO_2.

2. Principl

X-ray powder diffraction (XRD) is a rapid analytical technique primarily used for phase identification of a crystalline material, which can provide information on unit cell dimensions, etc. It is discovered that crystalline substances can act as three-dimensional diffraction gratings for X-ray wavelengths similar to the spacing of planes in a crystal lattice. Diffraction occurs when raying an X-ray beam on the crystalline sample. Then some diffraction peaks will become strong. In the simplest instance, an X-ray diffraction experiment consists of a set of diffracted intensities and the angles at which they are observed. This diffraction pattern can be thought of as a chemical fingerprint. Chemical identification can be performed by comparing this diffraction pattern to the known patterns from the database.

The Bragg equation, $2d\sin\theta = n\lambda$ is one of the keystones in understanding X-ray diffraction. In this equation, d is the interplanar spacing between rows of atoms, θ is the angle of the X-ray beam concerning these planes, n is an integer, and λ is the characteristic wavelength of the X-rays impinging on the crystallized sample. When this equation is satisfied, X-rays scattered by the atoms in the plane of a periodic structure are in phase and diffraction occurs in the direction defined by the angle θ.

From the diffraction pattern collected, chemical identification can be performed:

✓ Crystalline phases. It is the most important part of the XRD analytical technique. The crystalline phase of the testing sample can be identified by comparing the collected diffraction pattern to the known patterns in the database from the Joint Committee on Powder Diffraction Standards (JCPDS). If the collected diffraction pattern matches well with a particular standard diffraction pattern, then the crystalline phase for the testing sample will be identified.

✓ Crystallinity. Crystallinity means crystal quality. If the crystal has high crystallinity, the crystals will be big; the diffraction peaks will be strong and sharp. Half the height width of the diffraction peaks approaches the natural width of the instrument itself. On the contrary, crystals will be small and defects exist. The diffraction peaks will be weak and dispersed. The poorer the crystallinity, the weaker for the diffraction. The diffraction peaks will become wider, and even disappeared in the background.

✓ Particle size. The Debye-Scherrer equation, $D = K\lambda/(\beta\cos\theta)$ is the most important equation for calculating the size of the crystals. In this equation, D is the average thickness in the vertical direction of the crystal face; K is Scherrer constant (if β is FMHM, $K = 0.89$. if β is integral height to width of the diffraction peak, $K = 1$); λ is the wavelength of X-ray; β is the half-high width of the diffraction peak of the sample and θ is diffraction angle (deg). It should be noted that it is just the average size of the testing crystals as the size of the testing crystals could not be the same.

Titanium dioxide (TiO_2), also known as titanium white, is a white solid or powder sample with a molecular weight of 79.83. It is a kind of white inorganic pigment, which has the advantages of non-toxic, the best opacity and radiancy. It is regarded as the best white pigment up to now. Titanium dioxide has good chemical stability and also strong adhesion, which makes it a broad application in the fields of coatings, plastics, papermaking, printing inks, chemical fibers, rubber, make-ups, etc.

TiO_2 has three different phases, Anatase, Rutile, and Brookite. Among the three phases, Brookite is hard to be prepared and also not stable. The rest two phases are common, which belong to the tetragonal system. The standard diffraction pattern and PDF cards for Anatase and Rutile are listed below.

✓ Data for Anatase:

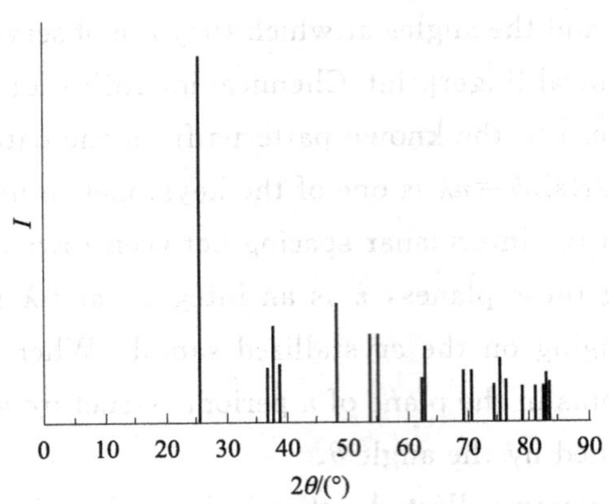

2θ	Int-f	h	k	l	2θ	Int-f	h	k	l	2θ	Int-f	h	k	l
25.325	999	1	0	1	62.165	27	2	1	3	78.754	1	2	0	6
36.989	62	1	0	3	62.750	118	2	0	4	80.860	4	0	0	8
37.841	190	0	0	4	68.842	51	1	1	6	82.236	6	3	0	3
38.601	72	1	1	2	70.346	55	2	2	0	82.759	43	2	2	4
48.074	249	2	0	0	74.155	5	1	0	7	83.232	18	3	1	2
53.952	158	1	0	5	75.129	85	2	1	5					
55.106	155	2	1	1	76.100	23	3	0	1					

Ref: Burdett, J. K et al., J. Am. Chem. Soc., 109, 3639, 1987.

✓ Data for Rutile:

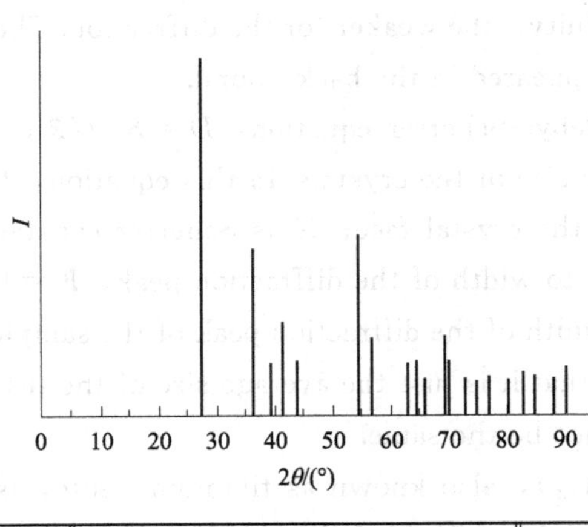

2θ	Int-f	h	k	l	2θ	Int-f	h	k	l	2θ	Int-f	h	k	l
27.459	999	1	1	0	62.799	60	0	0	2	76.597	13	2	0	2
36.104	399	1	0	1	64.105	56	3	1	0	79.894	8	2	1	2
39.223	54	2	0	0	65.565	5	2	2	1	82.414	24	3	2	1
41.269	184	1	1	1	69.064	140	3	0	1	84.333	17	4	0	0
44.081	64	2	1	0	69.851	67	1	1	2	87.568	7	4	1	0
54.364	444	2	1	1	72.483	8	3	1	1	89.628	40	2	2	2
56.676	131	2	2	0	74.470	2	3	2	0					

Ref: Sanchez, E et al., J. Solid State Chem., 122, 309, 1996.

In this experiment, the XRD test is performed for as-synthesized TiO_2 powers and commercial samples. According to the data collected, crystal phase analysis should be performed by comparing the diffraction pattern to the known patterns from the database.

3. Instrument and Reagents

3.1 Instrument:

XRD-Rigaku MiniFlex 600.

3.2 Reagents

Synthesized TiO_2 and commercial TiO_2 samples, alcohol.

4. Procedures

4.1 Take ~300mg dry powder sample on the holder. Press the sample to make it smooth and ready for the test.

4.2 Make the instrument ready for the test. Put the holder in the slot carefully and do the XRD test. Please set the file name, path, and measurement conditions (such as testing angles, steps, and speed).

4.3 Collect the samples after the test and then shut down the instrument.

5. Experimental HSE

Be aware of X-ray radiation.

6. Questions

6.1 What are the uses of TiO_2?

6.2 What information can be obtained from the X-powder diffraction pattern of the sample?

6.3 What are the crystalline phases of TiO_2 and how to deduce them from XRD data?

实验三 纳米二氧化钛光催化降解染料

一、实验目的

1. 了解纳米二氧化钛作为光催化剂的原理和应用。
2. 掌握光反应器的应用和光反应的原理。
3. 掌握纳米二氧化钛光催化降解染料的基本操作。

二、实验原理

印染行业是工业废水排放大户，据不完全统计，全国印染废水排放量为3～4百万立方米/天。染料废水是目前难降解的工业废水之一，对环境的危害日益严重。染料废水目前采用的处理方法主要有三种：物理法、化学法和生化法。物理法主要利用活性炭、硅藻土和煤渣等吸附处理，优点是费用低、脱色效果好，缺点是产生大量泥渣、进一步处理难。化学法主要是混凝沉淀法和混凝气浮法等，混凝法的工艺流程简单，设备投资少，缺点是运行费用高、泥渣量多。生化法是目前染料废水的主要处理方法，但是色度和COD去除率不高。

近年来，利用半导体粉末光催化处理难降解废水的研究日益引起人们的注意。本实验就是考察超细二氧化钛光催化降解染料的能力。

三、仪器与试剂

1. 仪器

NDC-3 型光学反应仪，紫外-可见分光光度计。

2. 试剂

超细二氧化钛（自制），亚甲基蓝溶液（30mg/L）。

四、实验步骤

1. 光降解操作

将 200mL 的某一浓度的染料亚甲基蓝溶液和 1.5％的超细二氧化钛催化剂加入光催化反应器中，开通冷却水，关闭暗箱门，开启汞灯，待汞灯工作稳定时开始计时，每隔 5 分钟取样 5mL 进行分析测定。

2. 染料降解率的测定

染料降解率按下式计算：

$$D=(A_0-A)/A_0\times100\%$$

式中，A_0、A 分别是光照前和光照 t 时间后染料溶液在最大吸收波长下的吸光度。

3. COD 值的测定

取一定量试液置于回流锥形瓶中，加入重铬酸钾标准溶液、硫酸-硫酸银溶液和硫酸汞，加热回流 2h（自开始沸腾时计时）。冷却后加入蒸馏水至溶液体积不少于 140mL。以试亚铁灵为指示剂，用硫酸亚铁铵标准溶液滴定，溶液颜色由黄色经蓝绿色至红褐色即为终点。

测定水样时，用蒸馏水按同样操作步骤做空白试验。

COD 值的变化用重铬酸钾法（COD_{Cr}）来测定：

$$COD_{Cr}(O_2,mg/L)=\frac{(V_0-V_1)\times c\times8\times1000}{V}$$

式中，c 为硫酸亚铁铵标准溶液的浓度，mol/L；V_0 为滴定空白时硫酸亚铁铵标准溶液的用量，mL；V 为滴定水样时硫酸亚铁铵标准溶液的用量，mL；8 为氧（O）摩尔质量，g/mol。

五、实验 HSE

1. 亚甲基蓝：可溶于水/乙醇，不溶于醚类。在空气中较稳定，其水溶液呈碱性，有毒。低毒，避免皮肤和眼睛接触。高浓度（5～10mg/kg；25～50mL 1％溶液）的亚甲蓝溶液对血红蛋白起氧化作用，使之生成高铁血红蛋白。该物质对环境可能有危害，对水体应给予特别注意。

2. 进行光催化降解实验时，请务必关闭暗箱门，以防紫外线照射。

3. 测试样品的反应池为石英材质，放入光催化反应器时请务必轻拿轻放。

4. 反应池中需要配套专用磁子进行搅拌操作。

六、思考题

1. 请查阅有关光催化降解的反应机理。

2. 探讨催化剂用量对光降解的影响。

3. 如何检验超细二氧化钛的重复应用性能？

Experiment 3 Photocatalytic degradation of dyes by Nano-TiO_2

1. Objectives

1.1 Understand the principle and application of nano-TiO_2 as photocatalyst.

1.2 Master the application of photoreactor and the principle of photoreaction.

1.3 Master the basic operation of photocatalytic degradation of dyes by nano-TiO_2.

2. Principle

A large amount of industrial wastewater was discharged by the printing and dyeing industry. According to incomplete statistics, the discharge of printing and dyeing wastewater in our country is 3 to 4 million m^3 per day. Dye wastewater is one of the current industrial wastewaters which hard to degrade. It is so harmful to the environment. There are three main treatment methods for dye wastewater: physical, chemical, and biochemical methods. The physical method is an adsorption treatment that uses activated carbon, diatomite, and cinder. Its advantages are low cost and good decolorization effect, but it will generate a huge amount of sludge and difficult to further process. The chemical method includes coagulation sedimentation and coagulation flotation. The coagulation process is simple and has a low equipment investment. The disadvantages are the operation will cost a lot and the generated amount of sludge is large. The biochemical method is currently the main treatment method for dye wastewater, but removal rates of the chroma and COD are not too high.

In recent years, the research on the photocatalytic treatment of wastewater by semiconductor powder has attracted more attention. The ability of superfine TiO_2's in photocatalytic degradation of dyes is tested in this experiment.

3. Instruments and Reagents

3.1 Instruments

NDC-3 optical reactor, UV-visible spectrophotometer.

3.2 Reagents

Self-made Nano-TiO_2, methylene blue solution (30mg/L).

4. Procedures

4.1 Photodegradation operation

200mL of a certain concentration of methylene blue solution and 1.5% nano-TiO_2 catalyst were added to the photocatalytic reactor, open the cooling water and mercury lamp after the dark box's door is closed. Starting to count when the mercury lamp is working stably, then sampling 5mL every 5 min for analysis and determination.

4.2 Determination of dye degradation rate

Dye degradation rate is calculated by the following formula:

$$D=(A_0-A)/A_0\times100\%$$

A_0 and A are the absorption values of the dye solution at the maximum absorption wavelength before and after the light, respectively.

4.3 Determination of COD value

Take a certain amount of test solution in a reflux conical flask and add potassium dichromate standard solution, sulfuric acid-silver sulfate solution, and mercury sulfate. Reflux 2h (counting after the solution is boiling). After cooling, distilled water is added until the solution volume is not less than 140mL. Use ferroin indicator solution as an indicator, and titration with ferrous sulfate ammonia standard solution. This process ends when the color of the solution turns from yellow to blue-green, and at last turns to reddish-brown.

When testing the water sample, use the same operation steps to make a blank test with distilled water.

The change of COD value is determined by the method of potassium dichromate (COD_{Cr}):

$$COD_{Cr}(O_2,mg/L)=\frac{(V_0-V_1)\times c\times8\times1000}{V}$$

c—— The concentration of ferrous sulfate ammonia standard solution, mol/L;

V_0——The amount of ferrous sulfate ammonia standard solution when titrating blank sample, mL;

V——The amount of ferrous sulfate ammonia standard solution when titrating water sample, mL;

8——half of the molar mass of oxygen (O), g/mol.

5. Experimental HSE

5.1 Methylene blues: soluble in water/ethanol, insoluble in ethers. Methylene blue is relatively stable in the air, and its aqueous solution is alkaline and toxic. Low toxicity, avoid skin and eye contact. High concentration (5 ~ 10mg/kg; 25 ~ 50mL 1% solution) methylene blue solution oxidizes hemoglobin to methemoglobin. This substance may be hazardous to the environment, and special attention should be given to water bodies.

5.2 When performing photocatalytic degradation experiments, please be sure to close the dark box door to prevent ultraviolet radiation.

5.3 The reaction cell of the test sample is made of quartz. Be sure to handle it gently when putting it into the photocatalytic reactor.

5.4 A special magnet is required in the reaction tank for stirring operation.

6. Questions

6.1 Please look up the reaction mechanism for photocatalytic degradation.

6.2 Discuss the effect of catalyst dosage on photodegradation.

6.3 How to test the repeated application performance of superfine TiO_2?

实验四 乙酰二茂铁的制备

一、实验目的

1. 掌握 Friedel-Crafts 反应合成乙酰二茂铁的反应机理和方法。
2. 学习乙酰二茂铁的合成方法。
3. 学习用薄层色谱跟踪反应进程和检测产品纯度的方法。

二、实验原理

1. 二茂铁及其衍生物

二茂铁（又称二环戊二烯合铁、环戊二烯基铁），分子式为 Fe（C_5H_5）$_2$，是一种典型的过渡金属与茂环生成的具有芳香性的有机金属化合物，常温下为橙色固体；它由两个环戊二烯基负离子和一个二价铁离子键合而成，具有夹心型结构。二茂铁可作为火箭燃料的添加剂、汽油的抗爆剂、硅树脂和橡胶的防老剂及紫外线吸收剂等。

二茂铁具有典型的芳香化合物性质，但对氧化剂敏感，其茂环上能发生多种取代反应，特别是亲电取代反应（例如 Friedel-Crafts 反应），比苯的反应活性高，在茂环上引入各种取代基可以形成丰富多彩的二茂铁衍生物。二茂铁衍生物由于高稳定性、低毒性、亲油疏水性和氧化还原可逆性等可作为抗菌剂、调节植物生长剂、特殊疾病药物、液晶材料和非线性光学材料等。

二茂铁的酰基化衍生物是一类重要的中间体，由于催化剂和反应条件的不同，二茂铁与乙酸酐在以磷酸为催化剂时可制得乙酰二茂铁或者 1,1′-二乙酰二茂铁。与苯的衍生物相似，由于乙酰基的钝化作用，两个乙酰基并不在同一茂环上。其反应式如下：

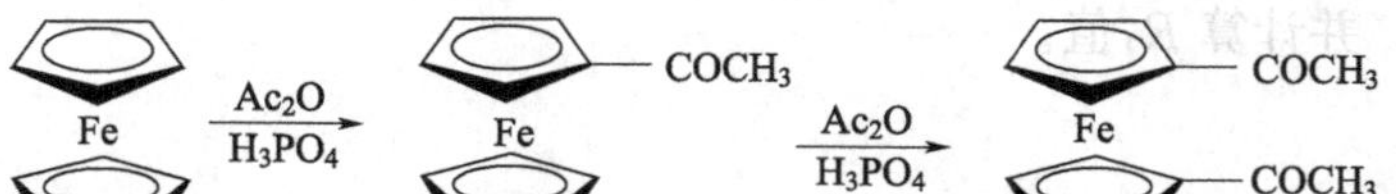

2. 薄层色谱（TLC）

由于混合物中的各个组分对吸附剂（固定相）的吸附能力不同，当展开剂（流动相）流经吸附剂时，发生无数次吸附和解吸的过程，吸附弱的组分随流动相迅速向前移动，吸附强的组分滞留在后，由于各组分具有不同的移动速率，最终得以在固定相薄层上分离。其应用主要有：跟踪反应进程；鉴定少量有机混合物的组成；分离；寻找柱色谱的最佳分离条件等。

展开剂的选择：主要根据样品的极性、溶解度和吸附剂的活性等因素考虑。一般常用溶剂极性从小到大排列顺序为：石油醚，环己烷，四氯化碳，苯，二氯甲烷，乙醚，四氢呋喃，乙酸乙酯，丙酮，乙醇，甲醇，水，吡啶，乙酸等。根据样品的不同，可使用单一溶剂，也可以使用混合溶剂。常用的溶剂组合有：石油醚/乙酸乙酯，石油醚/丙酮，石油醚/乙醚，石油醚/二氯甲烷，乙酸乙酯/甲醇，三氯甲烷/乙酸乙酯等。展开剂选择的条件包括：1）对待分离的组分有良好的溶解性；2）待分离组分的 R_f 在 0.2～0.8 之间，定量测定在

0.3～0.5之间；3）不与待分离组分或吸附剂发生化学反应；4）沸点适中，黏度较小；5）展开后组分斑点圆且集中；6）混合溶剂最好现用现配。

化合物R_f值：一个化合物在薄层色谱板上上升的高度与展开剂上升高度的比值称为化合物的R_f值，R_f值越大，该物质极性越小；反之则物质的极性越大。本实验中二茂铁、乙酰二茂铁和1,1-二乙酰基二茂铁在薄层色谱板上从上往下应出现黄色、橙色和红色三个点。

三、仪器与试剂

1. 仪器

磁力搅拌器，滴管，干燥管，沸水浴装置，电热套，烧杯500mL，圆底烧瓶（100mL），抽滤瓶，布氏漏斗，层析缸，紫外灯。

2. 试剂

二茂铁，磷酸（85%），乙酸酐，碳酸氢钠，石油醚，乙酸乙酯，薄层色谱硅胶板。

四、实验步骤

1. 乙酰二茂铁的制备

在100mL圆底烧瓶中，依次加入1g二茂铁和10mL乙酸酐，振荡下用滴管慢慢加入2mL 85%的磷酸（至深红色）。药品加入顺序不可颠倒且要控制磷酸的滴加速度。投料毕，用装有无水氯化钙的干燥管塞住瓶口，沸水浴上加热15min。将反应物倾入盛有40g碎冰的500mL烧杯中，用20mL冰水洗烧瓶，将洗液并入烧杯。在搅拌下，分批加入固体碳酸氢钠（至pH＝7～8），直到溶液呈中性为止，控制滴加速度防止产物溢出。将中和后的反应物置于冰浴中冷却15min，抽滤收集析出的橙黄色固体，压干后在空气中干燥，用石油醚（60～90℃）重结晶，熔点84～85℃，计算产率。

2. 乙酰二茂铁的薄层色谱（TLC）

取原料，反应混合物和粗产物溶于乙酸乙酯，分别点样在同一块色谱板上，用石油醚-乙酸乙酯（体积比为10∶1）的混合溶剂作为展开剂，用紫外灯观察斑点展开的结果，判别反应进行的情况，并计算R_f值。

五、实验HSE

1. 二茂铁：有特殊气味，急性毒性，LD50：1320mg/kg（大鼠经口），大于3000mg/kg（大鼠经皮）；吞咽或吸入有害，长期吸入或反复接触可能损害（肝）器官，对水生生物有毒。避免接触皮肤和眼睛，避免形成粉尘和气溶胶。急救措施：如果吸入，请移步到新鲜空气处；如果皮肤接触，请用肥皂和大量水冲洗；如果眼睛接触，请用水冲洗；如误食，禁止催吐，切勿给失去知觉者喂食任何东西及用水漱口。上述情况中，如果情节严重请及时就医。储存在阴凉处，容器保持密闭，储存于干燥通风处，切勿靠近火源，与强氧化剂分开存放。

2. 乙酸酐：无色透明液体，有强烈的乙酸气味，有吸湿性。低毒，易燃，有腐蚀性，勿接触皮肤或眼睛，以防引起损伤，有催泪性。急性毒性，LD50：1780mg/kg（大鼠经口）；4000mg/kg（兔经皮）；LC50：1000ppm，1h（大鼠吸入）。重度刺激，其蒸气与空气形成爆炸性混合物，遇明火、高热能引起燃烧爆炸。与强氧化剂可发生反应。急救措施：皮肤接触，立即脱去污染的衣着，用大量流动清水冲洗至少15min，就医；眼睛接触，立即提起眼睑，用大量流动清水或生理盐水彻底冲洗至少 15min，就医；吸入，迅速脱离现场至空气新鲜处，保持呼吸道通畅，如出现呼吸困难应立即就医；误服入口立即就医。储存于阴凉、干燥、通风良好环境中。远离火种、热源。应与氧化剂、还原剂、酸类、碱类、活性金属粉末、醇类等分开存放，切忌混储。

3. 磷酸：白色固体，空气中易潮解，无强氧化性，无强腐蚀性，属于较为安全的酸。低毒类，LD50：1530mg/kg（大鼠经口）；2740mg/kg（兔经皮），有刺激性。接触时注意防止入眼，防止接触皮肤，防止入口。磷酸蒸气能引起鼻黏膜萎缩；对皮肤有相当强的腐蚀作用，可引起皮肤炎症性疾患；能造成全身中毒现象。急救措施：若有磷酸蒸气入眼，应立即用大量生理盐水冲洗，随后送医院救治；如不慎溅到皮肤，应立即用大量清水冲洗，把磷酸洗净后，一般可用红汞溶液或龙胆紫溶液涂抹患处，严重时应立即送医院诊治。储存时应低温干燥，与 H 发孔剂、碱类分开存放。

4. 碳酸氢钠：俗称小苏打，白色粉末或单斜晶结晶性粉末，熔点 270℃，无臭，无毒，受热易分解，潮湿空气中易潮解。LD50：4220mg/kg（大鼠经口）。储存于阴凉、干燥、通风的环境中。远离火种、热源。与氧化剂、酸类分开存放。

5. 石油醚：无色透明液体，有煤油气味。易燃，与皮肤接触有刺激性。毒性数据：LD50：40mg/kg（小鼠静脉）；LC50：3400ppm，4h（大鼠吸入）。其蒸气或雾对眼睛、黏膜和呼吸道有刺激性。中毒表现为有烧灼感、咳嗽、喘息、喉炎、气短、头痛、恶心和呕吐。该品可引起周围神经炎。对皮肤有强烈刺激性。急救措施：皮肤接触，立即脱去污染的衣着，用肥皂水和清水彻底冲洗皮肤，就医；眼睛接触，立即提起眼睑，用大量流动清水或生理盐水彻底冲洗至少 15min，就医；吸入，迅速脱离现场至空气新鲜处，保持呼吸道通畅，如呼吸困难给输氧，如呼吸停止立即进行人工呼吸，就医；食入，用水漱口，给饮牛奶或蛋清，就医。储存于阴凉、通风的环境中。远离火种、热源。应与氧化剂分开存放，切忌混储。禁止使用易产生火花的机械设备和工具。

6. 乙酸乙酯：无色澄清液体，有强烈的醚气味，有甜味，低毒性。急性毒性，LD50：4940mg/kg（兔经口）；LC50：5760mg/m^3，8h（大鼠吸入）；人吸入 2000ppm×60min，出现严重毒性反应；人吸入 800ppm，有病症；人吸入 400ppm，短时间内眼、鼻、喉有刺激。易燃，其蒸气与空气可形成爆炸性混合物。遇明火、高热能引起燃烧爆炸。与氧化剂接触会剧烈反应。急救措施：吸入，迅速脱离现场至新鲜空气处，保持呼吸道通畅，如呼吸困难给输氧，如呼吸停止立即进行人工呼吸，就医；误食，饮足量温水，催吐，就医；皮肤接触，脱去被污染衣着，用肥皂水和清水彻底冲洗皮肤；眼睛接触，提起眼睑，用流动清水或生理盐水冲洗，就医。本品属于一级易燃品，应贮于低温阴凉通风处，远离火种火源。应与氧化剂、酸碱类等分开存放。

六、思考题

1. 二茂铁乙酰化的机理是什么？
2. 二茂铁比苯容易发生亲电取代，为什么不能用混酸进行硝化？
3. 乙酰二茂铁制备中为什么要求在无水的条件下进行？
4. 碳酸氢钠中和主要消除什么物质？
5. 实验中采用薄层色谱监测反应进程和产物纯度的原理是什么？

Experiment 4 Preparation of Acetylferrocene

1. Objectives

1.1 Learn the principle and method of preparing acetylferrocene by Friedel-Crafts Acylation reaction.

1.2 Learn the preparation method of Acetylferrocene.

1.3 Learn how to monitor the reaction progress and identify product purity by the thin-layer chromatography method (TLC).

2. Principle

2.1 Ferrocene and its derivatives

Ferrocene (known as Biscyclopentadienyl iron or Cyclopentenylferrocene) is an aromatic organometallic compound consisting of typical transition metal and cyclopentadiene with the formula Fe $(C_5H_5)_2$. It is an orange solid at room temperature. Moreover, it is the prototypical metallocene formed by two cyclopentadienyl rings bound on opposite sides of a central iron atom, which is known as sandwich structural compounds. Ferrocene compounds are usually used as a rocket fuel additive, antiknock of petroleum, antioxidant of silicon resin or rubber, and ultraviolet absorbent, etc.

Ferrocene possesses typical chemical properties of aromatic compounds but it is sensitive to the oxidants. It undergoes many electrophilic substitutive reactions (for example, the Friedel-Crafts reaction) in cyclopentadienyl rings to produce kinds of its numerous derivatives, which indicates higher reactivity than benzene. These ferrocene derivatives are applied as antibacterial agents, plant growth regulators, a drug of special disease, liquid crystal materials, and nonlinear optical materials because of their excellent properties including high stability, low toxicity, hydrophobicity, and redox reversibility.

Among these derivatives, ferrocene acylation compounds are important chemical intermediates. Typically, the compounds including acetylferrocene and 1,1-diacetylferrocene are obtained by the reaction of ferrocene with acetic anhydride in the presence of phosphoric acid. Moreover, they are changed with the differences between the catalysts and reaction conditions. Two acetyl groups are not linked in the same cyclopentadienyl ring due to their passivation. The preparation of these compounds shows in the following:

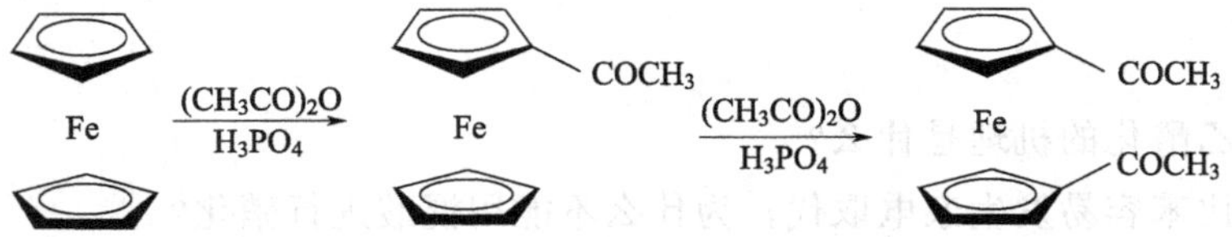

2.2 Thin-layer chromatography method (TLC)

The developing agent (known as the mobile phase) is selected according to the polarity of samples, the solubility of solvents, and the activity of adsorbents. The polarity of common solvents is arranged from low to high as the following order: petroleum ether, cyclohexane, carbon tetrachloride, benzene, dichloromethane, ethyl ether, tetrahydrofuran, ethyl acetate, acetone, ethanol, methanol, water, pyridine, acetic acid. The single solvent or mixed solvents are used as the developing agent according to the different analytes. Normally, the mixed solvents include the following components: petroleum ether/ethyl acetate, petroleum ether/acetone, petroleum ether/ether, petroleum ether/dichloromethane, ethyl acetate/methanol, chloroform/ethyl acetate, etc. The selecting criteria of developing agent includes the following aspects: 1) good solubility for the separated compounds; 2) the value of R_f is the range of 0.2～0.8, and the quantitative determination is the range of 0.3～0.8; 3) no reaction with the separated compounds or adsorbents; 4) moderate boiling point and low viscosity; 5) the developed spots are round and concentrated; 6) the mixed solvents are prepared and kept fresh when they are used.

The retardation factor R_f of compounds: the distance traveled by the substance being considered is divided by the total distance traveled by the mobile phase. The mobile phase must not be allowed to reach the end of the stationary phase. This ratio is called the retardation factor (R_f). In general, the compounds of high polarity show small R_f values, which is opposite to a large R_f value. In the experiment, the yellow, orange, and red spots are respectively denoted as ferrocene, acetyl ferrocene, and diacetyl ferrocene, which appear on the TLC plate from top to down by projecting ultraviolet light onto the plate.

3. Instruments and Reagents

3.1 Instruments

Magnetic stirrer, dropper, drying tube, boiling water bath, beaker, round-bottom flask, suction flask, Büchner funnel, Chromatography cylinder.

3.2 Reagents

Ferrocene, phosphoric acid (85%), acetic anhydride, sodium bicarbonate, petroleum ether, ethyl acetate, thin layer silica gel plate, Ultraviolet lamp.

4. Procedures

4.1 Preparation of Acetylferrocene: 1g ferrocene and 10mL acetic anhydride are successively added into a 100mL round-bottom flask. Then, 2mL phosphoric acid (85%) are slowly dropwise added into the above-mixed reactants under magnetic stirring. These reactants are added without reversal as the following order: ferrocene, acetic anhydride, phosphoric acid. Moreover, the dropping rate of phosphoric acid must be strictly controlled by stirring and slowly dropping with the dropper. After that, the bottleneck is sealed by the drying tube filled with anhydrous calcium chloride. The flask is placed on the boiling water bath for 15 minutes. Subsequently, the reactants are poured into a 500mL beaker with 40 g crushed ice. After the flask is washed by 20mL ice water, the washing liquor is added into the beaker, too. The mixtures in the beaker are neutralized by batch addition of sodium bicarbonate under stirring to pH = 7 ~ 8. Furthermore, when sodium bicarbonate is used to neutralize the crude product, you should add carefully to prevent the product from overflowing. Finally, the beaker with mixtures is cooled for 15 minutes in the ice bath. The orange-yellow solids are obtained through vacuum filtering, air drying, and petroleum ether recrystallization. The yield of acetylferrocene should be calculated and the melting point (about 84～85℃) should be detected.

4.2 Thin Layer Chromatography (TLC) of Acetylferrocene: The compounds including ferrocene, mixed reactants, and crude products are dissolved into the solution of ethyl acetate, respectively. Three samples are applied to the same TLC G plate. The mixed solvents of petroleum ether and ethyl acetate ($V/V = 10:1$) are used as a developer (that is mobile phase). After the experiment, the spots are visualized by projecting ultraviolet light onto the TLC plate. The reaction progress is concluded by TLC results. The retardation factors (R_f) are calculated according to the distance traveled by the substances.

5. Experimental HSE

5.1 Ferrocene: special odor. Toxicity: LD50: 1320mg/kg Oral (rat), $>$3000mg/kg Dermal (rat); It is harmful if swallowed or inhaled. Prolonged inhalation or repeated exposure may damage organs and be toxic to aquatic organisms. Avoid contact with skin and eyes, and avoid dust and aerosols. First-aid measures: if inhaled, move to fresh air; If skin contact occurs, rinse with soap and plenty of water. In case of eye contact, rinse with water; In case of accidental ingestion, do not induce vomiting. Do not feed anything and rinse with water to the unconscious person. In the above cases, if the circumstances are serious, please contact a doctor at once. Store in the shade. Keep the container closed in a dry and ventilated place. Keep away from fire.

5.2 Acetic anhydride: colorless, transparent liquid with a strong odor of acetic acid and hygroscopicity. Low toxic, flammable, corrosive, do not contact with skin or eyes to avoid damage, tearing property. Toxicity: LD50: 1780mg/kg Oral (rat); 4000mg/kg Dermal (rabbit); LC50: 1000 ppm, 1h Inhalational (rat). Severe irritation, its steam and air form an explosive mixture, combustion explosion happens when contacts open fire and high heat. Reacts with a strong oxidant. First-aid measures: skin contact, immediately take off contaminated clothing, rinse with plenty of flowing water for at least 15 minutes. Get medical aid; In case of eye contact, immediately lift eyelids, rinse thoroughly with plenty of flowing water or normal saline for at least 15 minutes, and get medical aid; Inhalation, quickly leave the scene to the fresh air, to maintain airway patency, if breathing difficulties should be treated immediately by seeking a doctor; Immediately seek medical treatment after ingestion. Store in a cool, dry, well-ventilated environment. Keep away from fire and heat. It should be stored separately with an oxidant, reducing agent, acid, alkali, active metal powder, alcohol, and so on. Try to avoid mixing storage.

5.3 Phosphoric acid: white solid, easy to moisture in the air, no strong oxidation, no strong corrosion, belong to the relatively safe acid, low toxicity, LD50: 1530mg/kg Oral (rat); 2740mg/kg Dermal (rabbit), irritation. Avoid contact with eyes, skin, and inhalation. Phosphoric acid vapor can cause atrophy of nasal mucosa. Producing a quite strong corrosive effect on the skin causes the skin inflammation disease; cause systemic poisoning. First-aid measures: if there is phosphoric acid vapor into the eyes, washed with a large amount of normal saline immediately, then sent to the hospital for treatment; If accidentally splash to the skin, rinse with a lot of water immediately, wash the phosphoric acid, generally available solution or gentian purple solution daub affected place, if the situation becomes serious, sent to the hospital immediately diagnosis and treatment. Storage should be dry at low temperature, and H hole agents, alkali separate storage.

5.4 Sodium bicarbonate: commonly known as baking soda, white powder or monocline crystalline powder, melting point 270℃, odorless, non-toxic, easy to decompose by heat, moisture in the air. LD50: 4220mg/kg Oral (rat). Store in a cool, dry, ventilated environment. Keep away from fire and heat. Store separately from oxidants and acids.

5.5 Petroleum ether: colorless transparent liquid with a smell of kerosene. Flammable, irritating in contact with skin, toxicity data: LD50: 40mg/kg vein (mouse); LC50: 3400ppm, 4h Inhalational (rat). Its vapor or mist can produce eye, mucous membrane, and respiratory irritations. Toxic manifestations may include a burning sensation, cough, wheezing, laryngitis, shortness of breath, headache, nausea, and vomiting. It can cause peripheral neuritis. A strong irritant to the skin. First-aid measures: skin contact, immediately take off the contaminated clothing, with soap and water thoroughly rinse the skin. Get medical aid. Eye contact: immediately lift eyelids and rinse thoroughly with plenty of running water or saline for at least 15 minutes. Get medical aid. Inhalation: quickly remove from the site to fresh air. Keep your airway open. If breathing is difficult, give oxygen. If breathing stops, give artificial respiration immediately. Get medical aid. Ingestion: rinse with water and give milk or egg whites to drink. Get medical aid. Store in a cool, ventilated environment. Keep away from fire and heat. Should be stored separately with an oxidant, avoid mixed storage. Do not use mechanical equipment and tools that are prone to spark.

5.6 Ethyl acetate: colorless clarifying liquid with a strong ether-like odor, sweet taste, low toxicity. Acute toxicity: LD50: 4940mg/kg Oral (rabbit); LC50: 5760mg/m^3, 8h Inhalational (rat), human inhalation of 2000 ppm×60min, severe toxic reaction; People who inhale 800ppm have symptoms; Human inhalation of 400 ppm for a short time, eyes, nose, throat irritation. Flammable, its vapor and air can form an explosive mixture. In the case of an open flame, high thermal energy causes a combustion explosion. It can cause a violent reaction when contact with an oxidant. First-aid measures: inhalation: quickly leave the scene to the fresh air. Keep your airway clean. If breathing is difficult, give oxygen. If breathing stops, give artificial respiration immediately and get medical aid. Ingestion: drink enough warm water, induce vomiting, and seek medical advice. Skin contact: remove contaminated clothing and rinse thoroughly with soap and water. Eye contact: lift eyelids and rinse with running water or normal saline. Get medical aid. This product belongs to the first-class inflammable product, should be stored in a cool and ventilated place at low temperature, away from the ignition source. It should be stored separately from oxidants, acids, and bases.

6. Questions

6.1 What is the mechanism of ferrocene acetylation?

6.2 Ferrocene is more susceptible to electrophilic substitution then benzene. Why can't we nitrify it with mixed acids?

6.3 Why is acetylferrocene prepared under an anhydrous condition?

6.4 What is removed in the neutralization process of sodium bicarbonate?

6.5 What are the principles of monitoring the reaction process and detecting product purity by TLC?

实验五 乙酰二茂铁粗产物的重结晶及红外光谱测定

一、实验目的

1. 了解重结晶的原理，熟悉重结晶提纯的操作。
2. 学习用红外吸收光谱进行化合物的定性分析。
3. 掌握用压片法制作固体试样晶片的方法。
4. 熟悉红外分光光度仪的工作原理及其使用方法。

二、实验原理

从有机合成反应分离出来的固体粗产物往往含有未反应的原料、副产物及杂质，必须加以分离纯化，重结晶是分离提纯固体化合物的一种重要的、常用的分离方法之一。重结晶利用混合物中各组分在某种溶剂中的溶解度不同，或在同一溶剂中不同温度时的溶解度不同，而使它们相互分离。固体有机物在溶剂中的溶解度受温度的影响较大，通常温度升高，溶解度增大；反之，则溶解度降低。对于前一种常见的情况，加热使溶质溶解于溶剂中，当温度降低时，其溶解度下降，溶液变成过饱和，从而析出结晶。由于被提纯化合物及杂质的溶解度不同，可以分离纯化所需物质。重结晶提纯法的一般过程包括选择溶剂、溶解固体、趁热过滤、去除杂质、晶体的析出。一般重结晶只适用于纯化杂质含量在5%以下的固体有机物。

红外吸收光谱：当一定频率（一定能量）的红外光照射分子时，如果分子某个基团的振动频率和外界红外辐射频率一致，二者就会产生共振。此时，光的能量通过分子偶极矩的变化传递给分子，这个基团就吸收一定频率的红外光，产生振动跃迁（由原来的基态跃迁到了较高的振动能级），从而产生红外吸收光谱。如果红外光的振动频率和分子中各基团的振动频率不一致，该部分红外光就不会被吸收。用连续改变频率的红外光照射某试样，将分子吸收红外光的情况用仪器记录下来，就得到试样的红外吸收光谱图。振动能级的跃迁伴随有转动能级的跃迁，因此所得的红外光谱不是简单的吸收线，而是一个个吸收带。在化合物分子中，具有相同化学键的原子基团，其基本振动频率吸收峰（简称基频峰）基本上出现在同一频率区域内，例如，$CH_3(CH_2)_5CH_3$、$CH_3(CH_2)_4C\equiv N$和$CH_3(CH_2)_5CH=CH_2$等分子中都有$-CH_3$、$-CH_2-$基团，它们的伸缩振动基频峰都出现在同一频率区域内，即在$3000cm^{-1}$波数附近，但又有所不同，这是因为同一类型原子基团，在不同化合物分子中所处的化学环境不同，使基频峰频率发生一定移动，例如$C=O$基团的伸缩振动基频峰一般出现在$1850\sim1860cm^{-1}$范围内，当它位于酸酐中时，$C=O$为$1820\sim1750cm^{-1}$；在酯类中时，为 1750～1725cm^{-1}；在醛中时，为 1740～1720cm^{-1}；在酮类中时，为 1725～1710cm^{-1}；在与苯环共轭时，如乙酰苯中 C=O 为 1695～1680cm^{-1}，在酰胺中时，C=O 为 1650cm^{-1}。因此，掌握各种原子基团基频峰的频率及其位移规律，就可应用红外吸收光谱来确定有机化合物分子中存在的原子基团及其在分子结构中的相对位置。

三、仪器与试剂

1. 仪器

25mL 圆底烧瓶，磁力搅拌器，回流冷凝管，玻璃棒，加热套，滤纸，布氏漏斗，抽滤瓶，真空泵，傅里叶红外光谱仪，压片机，磨具，干燥器，玛瑙研钵，药匙，擦镜纸，红外灯。

2. 试剂

石油醚（60～90℃），活性炭，沸石，光谱纯 KBr 粉末。

四、实验步骤

1. 乙酰二茂铁粗产物的重结晶

（1）称取 1g 乙酰二茂铁粗产物，放入 25mL 圆底烧瓶中，加入 5～6mL 石油醚和 2 粒沸石，加热至微沸，加热过程中用磁子不断搅拌，直至乙酰二茂铁完全溶解。稍冷，加入 1～2g 活性炭于溶液中，再次煮沸 5～10min，再补加 1mL 石油醚，煮沸。注意：活性炭不能加到沸腾的溶剂中，防止外溢。

（2）将溶解好的乙酰二茂铁溶液趁热抽滤。

（3）滤液静置，冷却，即有结晶析出。

（4）抽滤，用空心塞压挤晶体，将晶体抽滤至干。

（5）称重，计算产率。

2. 乙酰二茂铁的红外光谱测定

（1）固体样品的制备（溴化钾压片）

取 1～2mg 乙酰二茂铁置于玛瑙研钵中，加入已研细的大约 200mg 无水溴化钾，研磨成极细的粉末置于模具中，用压片机压成锭片。注意：在红外灯下研磨，尽量不要吸收水分，同时要防止烫伤。

（2）测绘乙酰二茂铁的红外吸收光谱

将锭片放在红外光谱仪的支架上，以空气为参比，记录红外光谱，并打印。

（3）分析乙酰二茂铁的红外光谱图

五、实验 HSE

石油醚是无色透明液体，有煤油气味。不溶于水，溶于无水乙醇、苯、氯仿、油类等多数有机溶剂。易燃易爆，与氧化剂可强烈反应。急性毒性，LD50：40mg/kg（小鼠静脉），LC50：3400ppm，4h（大鼠吸入）。急救措施：皮肤接触，立即脱去污染的衣着，用肥皂水和清水彻底冲洗皮肤；眼睛接触，立即提起眼睑，用大量流动清水或生理盐水彻底冲洗至少15min；吸入，则迅速脱离现场至空气新鲜处，保持呼吸道通畅，如呼吸困难给输氧，如呼吸停止立即进行人工呼吸；食入，则用水漱口，给饮牛奶或蛋清，就医。

六、思考题

1. 重结晶法主要包括几个步骤？各个步骤的主要目的是什么？

2. 某一有机物进行重结晶，最适合的溶剂应当具有哪些性质？

3. 加热溶解待重结晶的粗产物时，为什么加入溶剂的量比计算所需量略少？为什么要逐渐加入直至完全溶解？

4. 用活性炭脱色时，为什么要待固体物质完全溶解以后才加入活性炭？为什么不能在溶液沸腾的时候加入？

5. 用压片法制样时，为什么要求将固体样品试样研磨到颗粒粒度在 2μm 左右？为什么要求 KBr 粉末干燥？

6. 利用标准谱图进行化合物鉴定时要注意什么？

7. 红外谱图解析的一般过程是什么？

Experiment 5 Recrystallization of crude products of acetylferrocene and its characterization by infrared spectroscopy

1. Objectives

1.1 Understand the principle of recrystallizing and be familiar with the operation of recrystallizing purification.

1.2 Study the qualitative analysis of compounds by infrared absorption spectrum.

1.3 Master the method of making a solid sample wafer by pressing method.

1.4 Proficient in the working principle and usage of infrared spectrophotometer.

2. Principle

Solid crude products separated from organic synthesis reactions often contain unreacted raw material, by-products, and impurities, which must be separated and purified. Recrystallization is one of the most important separation methods for the separation and purification of solid compounds. Recrystallization is the separation of components of a mixture by their different solubility in a certain solvent or at different temperatures in the same solvent. The solubility of the solid organic compounds in the solvent is usually increased with the temperature increased. However, the solubility of some organic compounds is reduced. The solute is dissolved in solvents by heating and separated crystals when the solution becomes supersaturated with the decrease of temperature. Therefore, the compounds can be separated and purified based on the different solubility of the purified compounds and impurities. The recrystallization purification includes selecting a solvent, dissolving solids, filtering hot solution, removing the precipitation, and separating crystals. The recrystallization purification is only suitable for the purification of solid organic compounds with an impurity level of less than 5%.

When the vibration frequency of a certain group of the molecule is consistent with the external infrared radiation frequency, they will resonate. At this time, the energy of the light is transmitted to the molecule through the change of the molecular dipole moment. This group absorbs the infrared light of a certain frequency and generates a vibrational transition (from the original ground state to a higher vibrational level), thereby generates an infrared absorption spectrum. If the vibration frequency of the infrared light does not coincide with the vibration frequency of each group in the molecule, the portion of the infrared light will not be absorbed. An infrared absorption spectrum of the sample is obtained by irradiating a sample with infrared light whose frequency is continuously changed and recording the infrared light of the molecule by an instrument. Since the transition of the vibrational energy level is accompanied by a transition of the rotational energy level, the resulting infrared spectrum is not a simple absorption line but an absorption band. In a compound molecule, the fundamental vibration frequency absorption peak (abbreviated as a fundamental frequency peak) of the atomic group having the same chemical bond appears substantially in the same frequency region. For example, $CH_3(CH_2)_5CH_3$、$CH_3(CH_2)_4C{\equiv}N$ and $CH_3(CH_2)_5CH{=}CH_2$ and other molecules have $—CH_3$, $—CH_2—$ groups, and their stretching vibration fundamental frequency peaks appear in the same frequency region which is near $3000cm^{-1}$ wave number. However, the same type of atomic group has different chemical environments in different compound molecules, which causes the fundamental frequency peak frequency to move. For example, the fundamental frequency of the stretching vibration of the $C{=}O$ group generally appears in the range of $1850\sim1860cm^{-1}$. In the acid anhydride, $C{=}O$ is $1820\sim1750cm^{-1}$. In the ester, it is $1750\sim1725cm^{-1}$. In the aldehyde, it is $1740\sim1720cm^{-1}$. In the ketones, it is $1725\sim1710\ cm^{-1}$. In the conjugate with the benzene ring, such as acetonitrile, $C{=}O$ is $1695\sim1680cm^{-1}$. In the amide, $C{=}O$ is $1650cm^{-1}$, and so on. Therefore, by grasping the frequency and displacement law of various atomic groups, the infrared absorption spectrum can be used to determine the atomic groups present in the molecules of organic compounds and their relative positions in the molecular structure.

3. Instruments and Reagents

3.1 Instruments

25mL round bottom flask, magnetic stirrer, reflux condensing tube, glass rod, electric furnace, filter paper, Büchner funnel, suction filter bottle, vacuum pump, Fourier infrared spectrometer, tablet press, abrasive, dryer, agate mortar, medicine spoon, mirror paper, and an infrared lamp.

3.2 Reagents

Petroleum ether (60～90℃), activated carbon, zeolite, spectral pure KBr powder.

4. Procedures

4.1 Recrystallization of crude products of acetyl ferrocene

4.1.1 Weigh 1 g crude product of acetylferrocene, put it into a 25mL round-bottomed flask, add 5～6mL petroleum ether and two zeolites, then heat it to faint boiling point. During the heating process, keep stirring until acetylferrocene is completely dissolved. Add 1～2 g activated carbon in the solution when it is slightly cold, boil again for 5～10 minutes, add 1mL petroleum ether and boil. The activated carbon should not be added to the boiling solvent.

4.1.2 Filter the dissolved acetylferrocene solution fast with a Buchner funnel.

4.1.3 The filtrate is allowed to stand and cool, and crystal precipitation can be seen.

4.1.4 Squeeze the crystal with a hollow plug and drain the crystal until dry.

4.1.5 Weigh and calculate yield.

4.2 Characterization of acetylferrocene by infrared spectroscopy

4.2.1 Preparation of solid samples (potassium bromide tablet)

Take 1～2mg of acetylferrocene and place it in an agate mortar, add about 200mg of anhydrous potassium bromide, which is grinded into a superfine powder and place it in the mold, then press it into ingot tablets with tablet press.

4.2.2 Mapping the infrared absorption spectrum of acetylferrocene

Put the ingot tablets on the bracket of the infrared spectrometer, take air as a reference, record the infrared spectrum, then print.

4.2.3 Analyze the infrared spectra of acetylferrocene

5. Experimental HSE

Petroleum ether is a colorless and transparent liquid with the smell of kerosene. Petroleum ether is soluble in anhydrous ethanol, benzene, chloroform, oil, and other organic solvents, but not in water. Flammable and explosive, strong reaction with oxidant. Acute toxicity, LD50: 40mg/kg (mouse vein), LC50: 3400 ppm 4 hours (rat inhalation). The gas is irritating to the eyes, mucous membrane, and respiratory tract. First-aid measures: take off the contaminated clothes immediately in case of skin contact, and wash the skin thoroughly with soapy water and clear water. In case of eye contact, lift the eyelid immediately and wash thoroughly with a large amount of flowing water or normal saline for at least 15 minutes. If inhaled, quickly leave the site to the fresh air place. Keep the respiratory tract unobstructed. If breathing is difficult, give oxygen. If breathing stops, perform artificial respiration immediately. Rinse mouth with water and drink milk or egg white. Get medical attention.

6. Questions

6.1 How many steps are involved in recrystallization? What is the main purpose of each step?

6.2 What properties should be the most suitable solvent for the recrystallization of an organic substance?

6.3 Why is the amount of solvent added slightly less than calculated when heating and dissolving crude products to be recrystallized? Why should we gradually join the solvent until it dissolves completely?

6.4 When decolorization with activated carbon, why should we add activated carbon until the solid material dissolve completely? Why can't you add it when it's boiling?

6.5 Why is it necessary to grind the solid sample to a particle size of about 2μm when the sheet pressing method is used? Why should KBr powder be dry to avoid moisture absorption?

6.6 What should be paid attention to when using a standard spectrum to identify compounds?

6.7 What is the general process of infrared spectrum analysis?

实验六 乙酰二茂铁的熔点和^{1}H NMR测定

一、实验目的

1. 掌握乙酰二茂铁熔点的测定方法。
2. 掌握核磁共振的原理与基本结构。
3. 学会核磁共振谱图的测定方法与分析。
4. 了解核磁共振在实验中的具体应用。

二、实验原理

1. 熔点的定义：当固体物质加热到一定温度时，会从固体转变为液体，此时的温度称为熔点。严格来说熔点是指在101.325kPa下固相和液相处于平衡时的温度。

2. 纯物质都有固定的熔点，固-液两相间的变化非常敏锐，从初熔到全熔的温度范围一般不超过0.5～1℃（液晶除外）。当混有杂质后，化合物的熔点会降低，熔程会增长。因此，通过测定熔点可以鉴别固体物质的纯度。

3. 核磁共振现象来源于原子核的自旋角动量在外加磁场作用下的进动。根据量子力学原理，原子核与电子一样，也具有自旋角动量，其自旋角动量的具体数值由原子核的自旋量子数决定，实验结果显示，不同类型的原子核，其自旋量子数也不同：质量数和质子数均为偶数的原子核，自旋量子数为0；质量数为奇数的原子核，自旋量子数为半整数；质量数为偶数，质子数为奇数的原子核，自旋量子数为整数。迄今为止，只有自旋量子数等于1/2的原子核，其核磁共振信号才能够被人们利用，常用的原子核有^{1}H、^{11}B、^{13}C、^{17}O、^{19}F、^{31}P、^{77}Se。

由于原子核携带电荷，当原子核自旋时，会产生一个磁矩，这一磁矩的方向与原子核的自旋方向相同，大小与原子核的自旋角动量成正比。将原子核置于外加磁场中，若原子核磁矩与外加磁场方向不同，则原子核磁矩会绕外磁场方向旋转，这一现象类似陀螺在旋转过程中转动轴的摆动，称为进动。进动具有能量，也具有一定的频率。

原子核进动的频率由外加磁场的强度和原子核本身的性质决定，也就是说，对于某一特定原子，在一定强度的外加磁场中，其原子核自旋进动的频率是固定不变的。

原子核发生进动的能量与磁场、原子核磁矩以及磁矩与磁场的夹角相关，根据量子力学原理，原子核磁矩与外加磁场之间的夹角并不是连续分布的，而是由原子核的磁量子数决定的，原子核磁矩的方向只能在这些磁量子数之间跳跃，而不能平滑地变化，这样就形成了一系列的能级。当原子核在外加磁场中接受其他来源的能量输入后，就会发生能级跃迁，也就是原子核磁矩与外加磁场的夹角会发生变化。这种能级跃迁是获取核磁共振信号的基础。

为了让原子核自旋的进动发生能级跃迁，需要为原子核提供跃迁所需要的能量，这一能量通常是由外加射频场来提供的。根据物理学原理，当外加射频场的频率与原子核自旋进动的频率相同时，射频场的能量才能够有效地被原子核吸收，为能级跃迁提供助力。因此某种特定的原子核，在给定的外加磁场中，只吸收某一特定频率射频场提供的能量，这样就形成了一个核磁共振信号。

三、仪器与试剂

1. 仪器

显微熔点仪，载玻片，镊子，研钵，核磁管（进口）若干支，药匙和称量纸。

2. 试剂

乙酰二茂铁，氘代氯仿若干。

四、实验步骤

1. 熔点测定

测定时取很少量的样品（不多于 0.1mg），放在载玻片上，盖上盖玻片，轻轻研磨，让样品形成很薄的一层。在显微镜下观察，样品最好是分散得很小的颗粒，能看到颗粒形状即可，颗粒太小不利于观察，颗粒太大测量不准，更不能形成一片。因为样品如果堆积在一起，导热不均匀，一方面会使熔点测不准，另一方面会使熔程变长。准备好样品后盖上热台上装配的玻璃片（防止挥发样品污染物镜），调整焦距后加热。通常载玻片比较厚，导热比较慢，可以使用两个盖玻片夹住样品。对于不挥发的样品可以只用一片盖玻片托住样品，而不在样品上加盖。热台上的玻璃片也可以不加，但建议加，可以保护物镜。重要的是每次测定要采用同样的方法，因为样品与热台加盖与否，测定的结果都是不同的，有时误差还相当大。加热时，开始升温速度可快一些，当达到预计熔点温度以下 10～15℃时将升温速度调到 1～2℃/min。如果是指针式的电压表，可以调到 50V，升温速度比较快，接近熔点时调到 25V 左右（现在有些熔点测定仪可以程序升温，可以事先设定好这些参数，比较方便），具体情况具体分析。当颗粒形状变圆或出现明显液滴时记录初熔点，视野内完全变成液体时记录终熔点。

2. 核磁测定

（1）准备样品

① 样品纯度：进行核磁测试的样品纯度一般应大于 95%，无铁屑、灰尘、滤纸毛等杂质。

② 氘代溶剂：本仪器配置仅适合做液体样品，要求待测样品在某种氘代溶剂中有良好的溶解性能。

③ 核磁管：直径 5mm，长度至少为 16cm，要求管内外壁干净，管壁无划痕破损（严禁样品管在仪器探头内发生断裂，一旦断裂将造成重大仪器故障，需支付高昂的维修费用）。

④ 样品溶解：将样品溶解在氘代溶剂里，制成均一、透明的溶液，装入核磁管中（注意：如有不溶物，可超声助溶并过滤后装入，悬浊液和乳浊液不能测试）。溶剂用量：0.6mL（高度 4cm）（溶剂太少不能正常匀场）；样品用量：进行 ^{1}H NMR、^{19}F NMR 及 ^{31}P NMR 谱测试时需样 5～20mg；^{13}C NMR 谱测试时需样 30～50mg（因 ^{13}C 天然丰度较小，增加样品量可以节约机时和测试费用）。

（2）实验过程

① 样品制备：根据样品性质，选择好氘代试剂，将其溶解后加入核磁管，盖上核磁管帽，将核磁管擦拭干净后，放入蓝色转子中，用量深量筒调节好转子位置，打开气流开关，即可放入磁场中心的样品腔中。

② 按照实验仪器的操作说明，进行锁场、调谐和匀场。

③ 设置合适的实验参数，采集信号，调整相应参数，确定化学位移，积分，得到^1H NMR谱图。

④ 处理并打印谱图。

五、实验室HSE

1. 严禁佩戴心脏起搏器和人工关节的人员入内。
2. 严禁携带电子设备和具有存储功能卡片的人员入内。
3. 实验室内禁止抽烟、吃喝。
4. 未经许可不得进入实验室，独立操作仪器前必须接受安全培训。
5. 按照操作手册操作仪器。

六、思考题

1. 熔点和熔程有什么区别？
2. 怎样通过NMR确定所合成的二茂铁的结构是正确的？

Experiment 6 Determination of melting point and ^1H NMR of ferrocene

1. Objectives

1.1 Master the melting point determination of ferrocene.

1.2 Master the principle and basic structure of nuclear magnetic resonance.

1.3 Learn how to measure and analyze NMR spectra.

1.4 Understand the specific application of nuclear magnetic resonance in experiments

2. Principle

2.1 Definition of a melting point: When a solid substance is heated to a certain temperature, it will change from a solid to a liquid, and the temperature at this time is called the melting point. Strictly speaking, it refers to the temperature at which the solid phase and the liquid phase are at equilibrium at 101.325kPa.

2.2 Pure substances have a fixed melting point, and the change between two phases of solid and liquid is very sharp. The temperature range from initial melting to full melting generally does not exceed 0.5～1℃ (except for liquid crystal). When impurities are mixed, the melting point of the compound decreases, and the melting range increases. Therefore, the purity of the solid matter can be identified by measuring the melting point.

2.3 The nuclear magnetic resonance phenomenon is derived from the precession of the spin angular momentum of the nucleus under the action of an external magnetic field. According to the principle of quantum mechanics, the nucleus has the same spin angular momentum as the electron. The specific value of the spin angular momentum is determined by the spin quantum number of the nucleus. The experimental results show that different types of nucleus spin quantum numbers are also different: quality the number and proton number are even nucleus, the spin quantum number is 0; the mass number is an odd nucleus, the spin quantum number is a half-integer; the mass number is even, the proton number is an odd nucleus, and the spin quantum number is an integer. So far, only the nucleus with a spin quantum number equal to 1/2 can be used by people. The nucleus that is often used by people is ^{1}H, ^{11}B, ^{13}C, ^{17}O, ^{19}F, ^{31}P, ^{77}Se.

Since the nucleus carries a charge, when the nucleus spins, a magnetic moment is generated by the spin. The direction of the magnetic moment is the same as the spin direction of the nucleus, and the magnitude is proportional to the spin angular momentum of the nucleus. The nucleus is placed in an external magnetic field. If the magnetic moment of the nucleus is different from the direction of the applied magnetic field, the magnetic moment of the nucleus will rotate around the direction of the external magnetic field. This phenomenon is similar to the oscillation of the rotating shaft of the gyro during the rotation, which is called precession. Precession has an energy and a certain frequency.

The frequency of the nucleus precession is determined by the strength of the applied magnetic field and the nature of the nucleus itself. That is, for a particular atom, the frequency of the nucleus spin precession is fixed in an applied magnetic field of a certain intensity.

The energy of the precession of the nucleus is related to the magnetic field, the magnetic moment of the nucleus, and the angle between the magnetic moment and the magnetic field. According to the principle of quantum mechanics, the angle between the magnetic moment of the nucleus and the applied magnetic field is not continuously distributed, but by the nucleus. The magnetic quantum number determines that the direction of the magnetic moment of the nucleus can only jump between these magnetic quantum numbers, but cannot change smoothly, thus forming a series of energy levels. When the nucleus receives energy input from other sources in an applied magnetic field, an energy level transition occurs, that is, the angle between the magnetic moment of the nucleus and the applied magnetic field changes. This energy level transition is the basis for acquiring nuclear magnetic resonance signals.

For an energy-level transition in the precession of nuclear spin, it is necessary to provide the energy required for the transition of the nucleus, which is usually provided by the addition of an RF field. According to the physics principle, when the frequency of the applied RF field is the same as the frequency of the pre-nuclear spin precession, the energy of the RF field can be effectively absorbed by the nucleus to assist in the energy level transition. Therefore, a particular atomic nucleus, in a given applied magnetic field, only absorbs the energy provided by the RF field of a particular frequency, thus forming a nuclear magnetic resonance signal.

3. Instruments and Reagents

3.1 Instruments

Micro melting point apparatus, slides, tweezers, mortar, nuclear magnetic tube (import), branch, measuring paper, and weighing paper.

3.2 Reagents

Ferrocene and deuterated chloroform.

4. Procedures

4.1 Determination of melting point

Take a very small amount of sample (not more than 0.1mg) on the slide, cover it with a coverslip, and gently grind it to form a thin layer. Observed under the microscope, the sample is preferably small particles that are scattered, and the shape of the particles can be seen. The particles are too small to be observed, and the particles are too large to be measured, and it is impossible to form a piece because the samples are stacked together. The heat conduction will be uneven, on the one hand, the melting point will not be measured, and on the other hand, the melting range will be lengthened. After preparing the sample, cover the glass piece assembled on the hot stage (to prevent the sample of the volatile sample from being evaporated), adjust the focal length, and heat it. Usually, the slides are thicker and have a slower thermal conductivity. Two coverslips can be used to hold the sample. For non-volatile samples, only one coverslip can be used to hold the sample without capping the sample. The glass piece on the hot stage can also be omitted, but it is recommended to add it to protect the objective lens. It is important to use the same method for each measurement because the results are different when the sample is capped or not capped, and the hot stage is capped or not capped, sometimes the error is quite large. When heating, the heating rate can be started faster. When the temperature reaches 10～15℃ below the expected melting temperature, the heating rate is adjusted to 1～2℃. If it is a pointer type voltmeter, it can be adjusted to 50V, and the heating rate is faster. When it is close to the melting point, it can be adjusted to about 25V (some melting point testers can be programmed, and these parameters can be set in advance, which is convenient), and the specific situation needs to be explored. The initial melting point is recorded when the shape of the particles is rounded or significant droplets appear, and the final melting point is recorded when the field of view becomes completely liquid.

4.2 Nuclear magnetic measurement

4.2.1 Prepare samples

(1) Sample purity: The purity of samples subjected to nuclear magnetic testing should generally be >95%, without impurities such as iron filings, dust, filter paper wool.

(2) Deuterated solvent: This instrument configuration is only suitable for liquid samples, which can to be tested with good solubility in deuterated solvents.

(3) Nuclear magnetic tube: 5 mm in diameter and at least 16 cm in length, the inner

and outer walls of the tube are required to be clean, and the tube wall is free from scratches and damage. The sample tube is strictly prohibited from breaking in the instrument probe. Once it breaks, it will cause major instrument failure and require high maintenance cost.

(4) Dissolution of the sample: Dissolve the sample in a deuterated solvent to make a uniform, transparent solution, and put it into the nucleus magnetic tube (Note: if there is an insoluble matter, it can be filtered, and then loaded, suspension and milk. Turbidity cannot be tested). Solvent dosage: 0.6mL (height 4 cm) (too little solvent cannot be properly shimmed); sample dosage: 5～20mg for ^{1}H NMR, ^{19}F NMR, and ^{31}P NMR spectrum; ^{13}C NMR spectrum test sample is 30～50mg (due to the small abundance of ^{13}C, increasing the amount of sample can save machine time and testing costs).

4.2.2 Experiment procedure

(1) Sample preparation: According to the nature of the sample, select the deuterated reagent, dissolve it, add it to the nuclear magnetic tube, cover the nuclear magnetic tube cap, wipe the nuclear magnetic tube clean, put it into the blue rotor, and adjust the rotor position with the deep measuring cylinder. Open the airflow switch and place it in the sample chamber in the center of the magnetic field.

(2) Perform lock field, tuning, and shimming according to the operating instructions of the experimental instrument.

(3) Set the appropriate experimental parameters, collect the signal, adjust the corresponding parameters, determine the chemical shift, and integrate to obtain the ^{1}H NMR spectrum.

(4) Process and print the spectrum.

5. NMR Lab HSE

5.1 No entrance for those who are fitted with a heart pacemaker or artificial joints.

5.2 No entrance for those who carry electronic instruments or magnetic cards, such as electronic watches, ID cards.

5.3 Eating, drinking, and smoking in the lab are also prohibited.

5.4 Only authorized persons are allowed to enter the laboratory working areas, and all new staff and students must receive safety training before they are allowed to run experiments individually.

5.5 Everyone is required to follow the operation manuals.

6. Questions

6.1 What's the difference between the melting point and the melting range?

6.2 How do you confirm that the structure of synthesized ferrocene is correct according to NMR spectra?

实验七　聚苯胺的电化学合成

一、实验目的

1. 掌握应用电聚合的方法在 Pt 载体上制备聚苯胺膜。

2. 掌握电化学工作站的使用方法。

二、实验原理

将Pt电极置于苯胺（PA）的盐酸溶液中，施加一个线性扫描电位，可以在Pt电极载体上电聚成一层聚苯胺膜（PPA）。苯胺的电聚合符合阳离子自由基的聚合机理：首先是苯胺单体被氧化成阳离子自由基，而后大部分阳离子自由基在电极上聚合形成了由N原子和芳环上C-4的碳原子通过头-尾偶合的二聚产物，该过程重复进行即可使得聚苯胺链不断生长，如下所示。

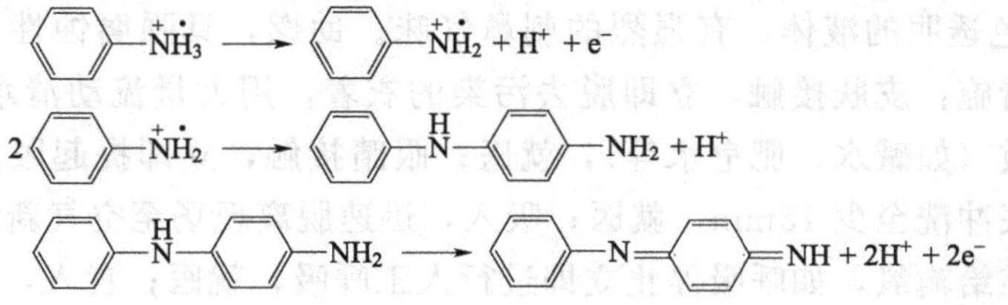

三、仪器与试剂

1. 仪器

电化学工作站，饱和甘汞电极，Pt盘电极，Pt丝电极，玻璃电解池，抛光布，抛光基座，超声清洗器，减压蒸馏装置一套，容量瓶（250mL），刻度移液管。

2. 试剂

苯胺（分析纯），盐酸（分析纯），二次蒸馏水，0.05μm氧化铝抛光粉。

四、实验步骤

1. 减压蒸馏苯胺

按有机化学实验操作，减压蒸馏苯胺，用油泵抽真空，蒸馏温度不得高于100℃，前馏分（约3～5mL）弃去，收集约10mL馏分，应为无色透明溶液。

2. 配制0.1mol/L苯胺＋1mol/L HCl溶液

用新鲜蒸馏的苯胺和盐酸配制250mL的0.1mol/L苯胺＋1mol/L HCl溶液。

3. Pt盘电极的前处理

Pt盘电极使用前抛光至镜面，然后用二次蒸馏水洗净，分别在乙醇、二次水中超声清洗备用。

4. 电解池的装配

将0.1mol/L苯胺＋1mol/L HCl溶液置入电解槽，以处理过的Pt盘电极为研究电极，饱和甘汞电极为参比电极，Pt丝电极为辅助电极组成三电极系统。

5. 电聚合

在电化学工作站控制软件中，选择CV方法，设定扫描电势范围－0.2～0.9V，扫描速度20mV/s，循环扫描20圈，记录CV曲线。

6. 后处理

实验结束后，取出Pt电极，用二次蒸馏水仔细冲洗电极表面，观察聚苯胺薄膜的特点。

五、实验HSE

1. 苯胺：无色或微黄色油状液体，有强烈气味。熔点为－6.2℃。遇明火、高热可燃。与酸类、卤素、醇类、胺类发生强烈反应，会引起燃烧。急性毒性，LD50：442mg/kg（大鼠-口服)，820mg/kg（兔-静脉）。急救措施：皮肤接触，立即脱去污染的衣着，用肥皂水和清水彻底冲洗皮肤，就医；眼睛接触，立即提起眼睑，用大量流动清水或生理盐水彻底冲洗至少15min，就医；吸入，迅速脱离现场至空气新鲜处，保持呼吸道通畅，如呼吸困难给输氧，如呼吸停止立即进行人工呼吸，就医；食入，饮足量温水，催吐，就医。储存注意事项：储存于阴凉、通风的库房。远离火种、热源。库温不超过30℃，相对湿度不超过80%。避光保存。包装要求密封，不可与空气接触。应与氧化剂、酸类、食用化学品分开存放，切忌混储。配备相应品种和数量的消防器材。储区应备有泄漏应急处理设备和合适的收容材料。

2. 盐酸：无色透明的液体，有强烈的刺鼻气味。助燃，具强腐蚀性、强刺激性，可致人体灼伤。急救措施：皮肤接触，立即脱去污染的衣着，用大量流动清水冲洗至少15min，可涂抹弱碱性物质（如碱水、肥皂水等），就医；眼睛接触，立即提起眼睑，用大量流动清水或生理盐水彻底冲洗至少15min，就医；吸入，迅速脱离现场至空气新鲜处，保持呼吸道通畅，如呼吸困难给输氧，如呼吸停止立即进行人工呼吸，就医；食入，用大量水漱口，吞服大量生鸡蛋清或牛奶（禁止服用小苏打等药品），就医。储存注意事项：储存于阴凉、通风的库房。库温不超过30℃，相对湿度不超过85%。保持容器密封。应与碱类、胺类、碱金属、易（可）燃物分开存放，切忌混储。储区应备有泄漏应急处理设备和合适的收容材料。

六、思考题

1. 为什么Pt电极在聚合前要认真地进行前处理？
2. 苯胺电聚合的机理是什么？

Experiment 7 Electrosynthesis of Polyaniline

1. Objectives

1.1 Grasp the method of electropolymerization to prepare polyaniline film on Pt.

2.2 Master the usage of the electrochemical workstation.

2. Principle

The Pt electrode is placed in a hydrochloric acid solution of aniline (PA) and a linear scanning potential is applied to form a polyaniline film (PPA) on the Pt electrode. A cationic polymerization mechanism has been proposed for the electropolymerization of PA: The cationic radicals are first generated from the oxidization of PA monomers, and then form the dimers on the electrode by the link between N atoms and C-4 atoms on aromatic rings. With successive coupling, PPA film is formed on the surface of the electrode.

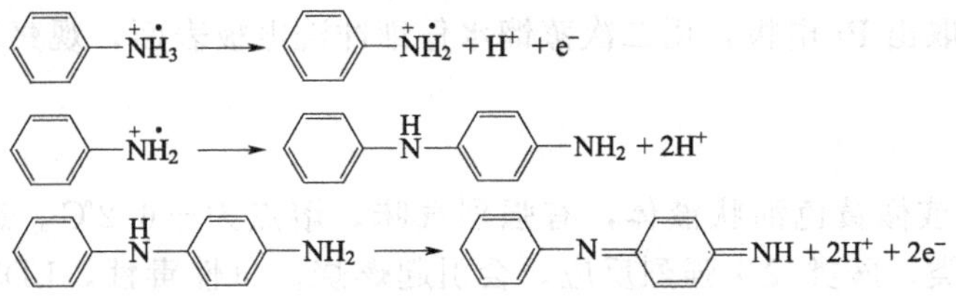

3. Instruments and Reagents

3.1 Instruments

Electrochemical workstation, saturated calomel electrode, Pt disk electrode, Pt wire electrode, glass electrolytic cell, polishing cloth, polishing base, ultrasonic cleaner, vacuum distillation device, a volumetric flask (250mL), graduated pipette.

3.2 Reagents

Aniline (analytically pure), hydrochloric acid (analytically pure), secondary distilled water, 0.05μm alumina polishing powder.

4. Procedures

4.1 Based on the operation of the organic chemistry experiment, aniline is vacuum distilled with an oil pump, and distillation temperature should not be higher than 100℃. During distillation, the initial fraction (3~5mL) should be discarded, and collect about 10mL fraction which is a colorless transparent solution.

4.2 Prepare 250mL of 0.1mol/L aniline+1mol/L HCl solution with freshly distilled aniline and hydrochloric acid.

4.3 Pretreatment of Pt Disk Electrode

The Pt disk electrode is polished to a mirror before use, then washed with distilled water, and ultrasonically washed in ethanol and finally, in bi-distilled water.

4.4 Assembly of an electrolytic cell

The solution containing 0.1mol/L aniline+1mol/L HCl was put into an electrolytic cell. The treated Pt disk electrode was used as the working electrode, the saturated calomel electrode was used as the reference electrode, and the Pt wire electrode was used as the auxiliary electrode to constitute a three-electrode system.

4.5 Electropolymerization

In the control software of the electrochemical workstation, the CV method is chosen with scanning potential range −0.2~0.9V, scanning speed 20mV/s and cyclic scanning 20 turns, and then CV curves are recorded.

4.6 Post-treatment

After the experiment, the Pt electrode was taken out, and the surface of the electrode was washed carefully with bi-distilled water to observe the characteristics of polyaniline film.

5. Experimental HSE

5.1 Aniline: colorless or slightly yellow oily liquid with a strong odor. The melting point is −6.2℃. Combustible in case of fire or high heat. Reacts violently with acids, halogens, alcohols, amines and will cause combustion. Acute toxicity, LD50: 442mg/kg (rat-oral); 820mg/kg (rabbit-venous). First-aid measures: Skin contact: take off contaminated clothing immediately and flush the skin thoroughly with soap water and water. See a doctor. Eye contact: lift the eyelid immediately and rinse thoroughly with plenty of running water or saline for at least 15 minutes. See a doctor. Inhalation: Quickly leave the scene to fresh air. Keep the airway clear. If breathing is difficult, give oxygen. If breathing stops, give artificial respiration immediately. See a doctor. Ingestion: Drink plenty of warm water to induce vomiting. See a doctor. Precautions for storage：Store in a cool，ventilated warehouse. Keep away from fire and heat. The storage temperature does not exceed 30℃ and the relative humidity does not exceed 80%. Keep away from light. The packaging needs to be sealed and not in contact with air. It should be stored separately from oxidants，acids，and edible chemicals，and avoid mixed storage. Equipped with the appropriate variety and quantity of fire equipment. The storage area should be equipped with leakage emergency treatment equipment and suitable containment materials.

5.2 Hydrochloric acid：It is a colorless and transparent liquid with a strong pungent odor and high corrosion. Combustion-supporting，highly corrosive and irritating，which can cause burns to the human body. First-aid measures：Skin contact：immediately remove contaminated clothing，flush the skin with plenty of flowing water for at least 15 minutes，apply weak alkaline substances（such as alkaline water，soapy water，etc.）and seek medical attention. Eye contact：lift the eyelid immediately and rinse thoroughly with plenty of running water or saline for at least 15 minutes. See a doctor. Inhalation：Quickly leave the scene to go out for some fresh air. Keep the airway clear. If breathing is difficult，give oxygen. If breathing stops，give artificial respiration immediately. See a doctor. Ingestion：Gargle with plenty of water，swallow a large number of raw egg whites or milk（do not take drugs such as baking soda），and seek medical attention. Precautions for storage：Store in a cool，ventilated warehouse. The storage temperature does not exceed 30℃ and the relative humidity does not exceed 85%. Keep the container tightly closed. It should be stored separately from alkalis，amines，alkali metals，and combustible materials. Avoid mixed storage. The storage area should be equipped with leakage emergency treatment equipment and suitable containment materials.

6. Questions

6.1 Why should Pt electrodes be carefully pretreated before polymerization?

6.2 What is the mechanism of aniline electropolymerization?

实验八 聚苯胺导电聚合物膜的pH响应

一、实验目的

1. 掌握聚苯胺膜对 pH 的响应原理。
2. 掌握循环伏安法研究电极反应的一般方法。

二、实验原理

酸性条件下电化学聚合的聚苯胺在不同 pH 下可发生质子化和去质子化形态转变，如下所示，这种性质使聚苯胺具有良好的 pH 响应性，在不同 pH 下表现出不同的电化学行为。

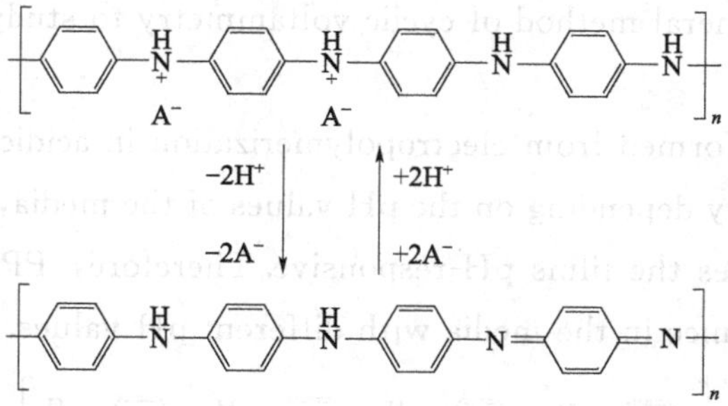

三、仪器与试剂

1. 仪器

电化学工作站，饱和甘汞电极，制备了聚苯胺膜的 Pt 盘电极，Pt 丝电极，玻璃电解池，容量瓶（250mL），刻度移液管。

2. 试剂

三种 pH 缓冲溶液（4.00，6.86，9.18）。

四、实验步骤

1. 电解池的装配

将 pH=4.00 的缓冲溶液置入电解槽，以制备了聚苯胺膜的 Pt 盘电极为研究电极，饱和甘汞电极为参比电极，Pt 丝电极为辅助电极组成三电极系统。

2. 循环伏安测试

在电化学工作站控制软件中，选择 CV 方法，设定扫描电势范围−0.2～0.9V，扫描速度 10mV/s，记录聚苯胺在 pH 缓冲溶液中的 CV 曲线；取出研究电极，用二次蒸馏水仔细清洗电极，并在二次蒸馏水中浸泡 10min，然后再依次测试 pH 为 6.86 和 9.18 的缓冲溶液中的循环伏安曲线。

五、实验 HSE

无。

六、思考题

1. pH 值为什么会影响聚苯胺的循环伏安行为？
2. 除了 pH 外，pH 缓冲溶液中还有哪些离子可能影响聚苯胺的循环伏安行为？

Experiment 8 pH response of polyaniline conductive polymer film

1. Objectives

1.1 Grasp the response of polyaniline film to pH value.

1.2 Master the general method of cyclic voltammetry to study electrode reactions.

2. Principle

For the PPA films formed from electropolymerization in acidic media, their structures and compositions can vary depending on the pH values of the media, as shown in the following scheme, which makes the films pH-responsive. Therefore, PPA films exhibit different electrochemical performance in the media with different pH values.

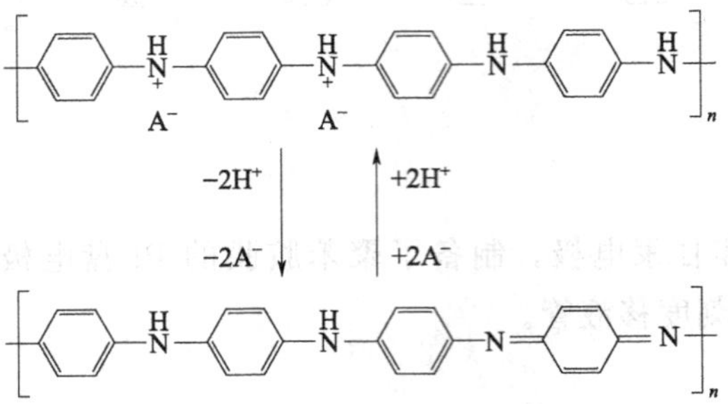

3. Instruments and Reagents

3.1 Instruments

Electrochemical workstation, saturated calomel electrode, Pt disk electrode with polyaniline film, Pt wire electrode, glass electrolytic cell, volumetric flask (250mL), graduated pipette.

3.2 Reagents

Three pH buffer solutions (4.00, 6.86, and 9.18).

4. Procedures

4.1 Assembly of an electrolytic cell

The buffer solution with pH=4.00 was put into an electrolytic cell. The Pt disk electrode with polyaniline film prepared was used as the research electrode, the saturated calomel electrode was used as the reference electrode, and the Pt wire electrode was used as the auxiliary electrode to form a three-electrode system.

4.2 Cyclic voltammetry test

In the control software of the electrochemical workstation, the CV method is chosen with a scanning potential range −0.2～0.9 V and scanning speed 10 mV/s, and then the CV curve of polyaniline in pH buffer solution is recorded. Take out the research electrode, carefully clean the electrode with bi-distilled water, soak it in bi-distilled water for 10 min, and then test the cyclic voltammograms in solutions with pH=6.86 and 9.18 in turn.

5. Experimental HSE

None.

6. Questions

6.1 Why does pH affect the cyclic voltammetric behavior of polyaniline?

6.2 In addition to pH, which ions in pH buffer solution may affect the cyclic voltammetric behavior of polyaniline?

实验九 2-氨基-1,3,4-噻二唑的制备

一、实验目的

1. 掌握1,3,4-噻二唑类化合物的合成反应机理和方法。
2. 学习一种酸催化制备杂环化合物的方法。
3. 巩固重结晶法纯化有机化合物的操作技能。

二、实验原理

1,3,4-噻二唑类化合物是一类重要的有机中间体，因其是含有硫、氮杂原子的五元杂环化合物，且具有芳香性及共轭效应，广泛用于合成抗菌药物、抗结核药物、抗癌药物、抗炎药物、除草剂和荧光传感分子等，而2-氨基-1,3,4-噻二唑因为2位上氨基的活泼性较高，容易将1,3,4-噻二唑引入到其他活性体系中获得更优的生物活性而备受关注。2-氨基-1,3,4-噻二唑为白色固体结晶，易溶于热水、乙醇，难溶于乙醚、氯仿、苯，熔程为190～192℃。

目前，噻二唑类化合物主要是在酸催化下由各种羧酸和硫代氨基脲闭环反应得到的。主要使用的酸催化剂有三氯氧磷、盐酸、硫酸和多聚磷酸等，催化剂为三氯氧磷时反应要求较高，操作复杂，浓盐酸因产率较高且价廉易得，是制备2-氨基-1,3,4-噻二唑最常用的催化剂。2-氨基-1,3,4-噻二唑类化合物合成过程监测比较困难，目前主要以延长反应时间、提高回流温度等方法获得较高产率。

具体可能的反应机理如下：首先是甲酸和氨基硫脲形成酰胺，进一步异构化为硫醇，再在盐酸催化下脱水，得到目标化合物。

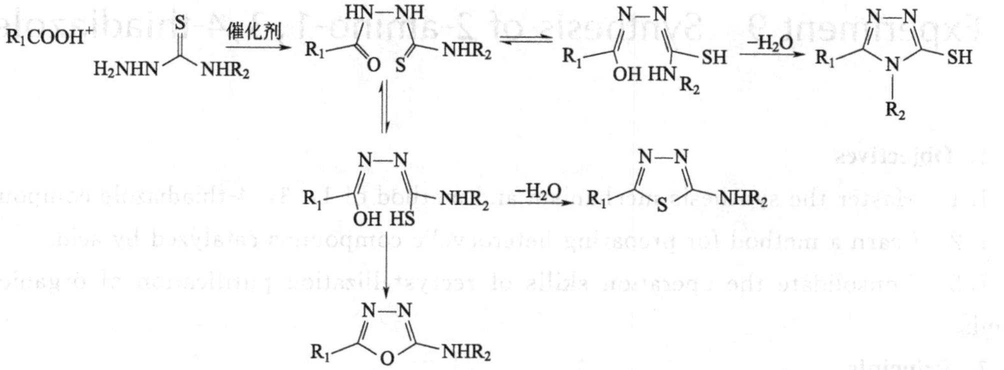

三、仪器与试剂

1. 仪器

磁力搅拌器，电热套，广泛pH试纸，布氏漏斗，抽滤瓶，真空泵，圆底烧瓶(100mL)，尾气吸收装置。

2. 药品

硫代氨基脲，盐酸（36%～38%），甲酸，氢氧化钠，乙醇。

四、实验步骤

在带有搅拌和回流冷凝器的 100mL 三口烧瓶中加入 0.088mol（8.00g）硫代氨基脲，0.104mol（5.4g）甲酸，然后向其中加入 50mL 浓盐酸，同时开启搅拌和加热，回流反应 2h，反应完成后冷却至室温。冰水浴下用 40%的氢氧化钠水溶液调 pH 至 7～8，室温冷却到有固体，再在冰水浴下静置 30min 以充分析出结晶，减压过滤，少量冰水洗涤后干燥得粗品。用 30%乙醇水溶液重结晶，干燥后得到 2-氨基-1,3,4-噻二唑纯品，测其熔点。

五、实验 HSE

硫代氨基脲：遇明火、高热可燃。与酸类、卤素、醇类、胺类发生强烈反应，会引起燃烧。急性毒性，LD50：442mg/kg（大鼠-口服）；820mg/kg（兔-静脉）。急救措施：皮肤接触，立即脱去污染的衣着，用肥皂水和清水彻底冲洗皮肤，就医；眼睛接触，立即提起眼睑，用大量流动清水或生理盐水彻底冲洗至少 15min，就医；吸入，迅速脱离现场至空气新鲜处，保持呼吸道通畅，如呼吸困难给输氧，如呼吸停止立即进行人工呼吸，就医；食入，饮足量温水，催吐，就医。储存注意事项：储存于阴凉、通风的库房。远离火种、热源。库温不超过 30℃，相对湿度不超过 80%。避光保存。包装要求密封，不可与空气接触。应与氧化剂、酸类、食用化学品分开存放，切忌混储。配备相应品种和数量的消防器材。储区应备有泄漏应急处理设备和合适的收容材料。

六、思考题

1. 该反应可能有哪些副产物，它们是如何产生的？
2. 如何判断所得产物是否为纯品？

Experiment 9 Synthesis of 2-amino-1,3,4-thiadiazole

1. Objectives

1.1 Master the synthesis mechanism and method of 1，3，4-thiadiazole compounds.

1.2 Learn a method for preparing heterocyclic compounds catalyzed by acid.

1.3 Consolidate the operation skills of recrystallization purification of organic compounds.

2. Principle

1,3,4-thiadiazole compounds are a kind of important organic intermediates containing sulfur，nitrogen heteroatom five-membered heterocyclic compounds，which have aromaticity and conjugation effect and widely used in the synthesis of antibacterial drugs，anti-tuberculosis drugs，anti-cancer drugs，anti-inflammatory drugs，herbicides，and fluorescent sensor molecules. 2-amino-1,3,4-thiadiazole is a major concern because of the high lively 2-amino and easy to be introduced into the other active system to gain better biological activity. 2-amino-1,3,4-thiamethoxam, as the white solid crystal, easily soluble in hot water, ethanol, but insoluble in ether, chloroform, benzene. The Melting point of 2-amino-1,3,4-thiamethoxam is 190～192℃.

The main synthesis method of thiadiazole compounds is obtained by the ring-closing reaction of various carboxylic acids and thiocarbazide under the catalysis of acid. Mainly using acid catalysts such as phosphorus oxychloride, hydrochloric acid, sulfuric acid, and polyphosphoric acid. Phosphorus oxychloride catalyst reaction is demand higher and operation complicated, while concentrated hydrochloric acid is the most common catalyst in the preparation of 2-amino-1,3,4-thiamethoxam because of higher yield and low cost. The azole compounds synthesis process monitoring of 2-amino-1, 3, 4-thiamethoxam is more difficult, at present mainly by prolonging the reaction time and reflux temperature to obtain a high yield rate.

The possible reaction mechanism is as follows: firstly, formic acid and thiosemicarbazide are formed into amides, which are further isomerized into mercaptan, and then dehydrated under the catalysis of hydrochloric acid to obtain the target compound.

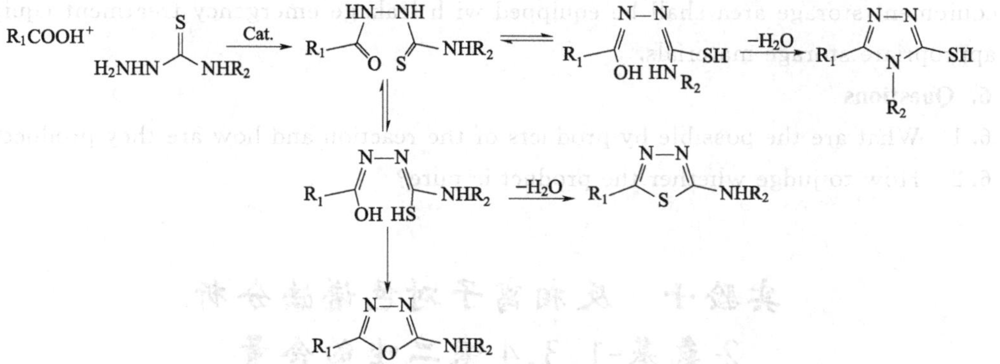

3. Instruments and Reagents

3.1 Instruments

Electromagnetic stirrer, heating mantle, pH-indicator paper, Büchner funnel, filter flask, vacuum pump, round bottom flask (100mL), and exhaust absorption device.

3.2 Reagents

Thiosemicarbazide, hydrochloric acid (36%～38%), formic acid, sodium hydroxide, ethanol.

4. Procedures

0.088mol (8.00g) thiosemicarbazone and 0.104mol (5.4g) formic acid were added into the 100mL 3-port round bottom flask with stirring and reflux condenser. Then, 50mL of concentrated hydrochloric acid was added to the flask. At the same time, stirring and heating were started. After refluxing for 2 h, the reaction was cooled to room temperature.

Under the ice water bath, the mixture was adjusted to pH=8 with a 40% sodium hydroxide solution, and the mixture was cooled to produce solid at room temperature and allowed to added 30 min in the ice water bath to fully precipitate crystallization. The white solid powder was obtained by filtration and washed with ice water，then recrystallized with 30% ethanol. Measure the melting point of this recrystallized 2-amino-1,3,4-thiamethoxam.

5. Experimental HSE

Hiocarbamide：When exposed to the high heat of an open flame，it can ignite and react strongly with the acid halide alcohols and amines，which will cause combustion. Acute toxicity，LD50：442mg/kg（rat-oral）；820mg/kg（rabbit-venous）. First aid measures：skin contact：immediately remove contaminated clothing and thoroughly flush the skin with soap and water and then go to see a doctor；eye contact：immediately filed an eyelid，flushed with plenty of liquid water or saline thoroughly at least 15 minutes，and then go to a doctor；inhalation：quickly from the scene to go out for some fresh air to keep respiratory tract unobstructed，give oxygen when it is difficulty in breathing，artificial respiration when the breathing is stop immediately，and then go to see a doctor；eating：drink enough water，vomiting，medical. Storage precautions：Store in a cool and ventilated warehouse away from the fire heat source temperature is not more than 30℃，the relative humidity shall not exceed 80% to avoid light，and the packaging shall be sealed. It shall not be in contact with air and shall be stored separately from an oxidizing agent，acid，and edible chemicals. The fire equipment storage area shall be equipped with leakage emergency treatment equipment and appropriate storage materials.

6. Questions

6.1 What are the possible by-products of the reaction and how are they produced?

6.2 How to judge whether the product is pure?

实验十　反相离子对色谱法分析 2-氨基-1,3,4 噻二唑的含量

一、实验目的

1. 掌握反相离子对色谱法的分离原理及其应用。
2. 巩固高效液相色谱的使用方法及注意事项。

二、实验原理

2-氨基-1,3,4 噻二唑（敌枯唑）对防治水稻白叶枯病等作物细菌性病害有很好的药效，但是，动物试验表明 2-氨基-1,3,4 噻二唑有强的致畸作用。在接触这类化合物的一些人员身上曾发现了不同程度的舌炎、角质层皮炎、过尿酸血症等，并在面部和外阴部出现红斑。环保监测中，需要对 2-氨基-1,3,4 噻二唑残留量进行分析。

敌枯唑的极性较强，又有一定的碱性。在硅胶柱上，选用氯仿-异丙醇-醋酸（40∶10∶1）为流动相，色谱峰有不对称情况，拖尾现象较为严重。在反相 ODS 柱上，用通常的甲醇-水或乙腈-水体系作为流动相，没有保留。薛慰灵等使用氯仿-乙醇-二甲基甲酰胺（95∶5∶

0.2，体积比）作为流动相，氨基键合相为固定相分离分析敌枯唑和敌枯双的含量。Ackerly等使用1-辛烷磺酸为离子对试剂，在C18柱上可以分离测定血浆中的敌枯唑。黄磊等也采用ODS柱，以含有0.005mol/L离子对试剂1-辛烷磺酸的稀甲醇溶液为流动相检测敌枯唑。

固定相极性小于流动相极性时的液相色谱分离模式称为反相色谱。流动相中加入离子对试剂后，它可以与被分析物（多为在溶液中带与离子对试剂相反的电荷）结合成离子对而呈电中性，这样非极性就表现出来，从而增加了在反相色谱柱上的保留作用。这种分离模式叫作反相离子对色谱。常用离子对试剂有两大类：四丁基铵磷酸盐和烷基磺酸的盐类。四丁基铵盐在pH=7.5的条件下与强酸和弱酸形成离子对并具有缓冲抑制弱碱的作用；多烷基磺酸盐在pH=3.5的条件下与强碱和弱碱形成离子对并缓冲抑制弱酸离子。在反相离子对色谱体系中影响流动相对溶质的保留和分离作用的因素主要有三个方面，即流动相组成、离子对试剂的浓度、有机改性剂。

反相离子对色谱常用的流动相是以水为主体的缓冲溶液，其pH值对分离的影响最大。因为流动相的酸碱度决定了被分析物的解离状态，从而影响其与离子对试剂的结合程度，因此，改变流动相的pH值是改善分离选择性的有效方法。离子对试剂的浓度是影响样品保留的主要因素，增加离子对试剂的浓度可以增加保留作用。以水为主体的流动相中通常添加极性有机溶剂如甲醇、乙腈等作为改性剂，其作用是改善缔合分子在两相中的分配，同时对某些化合物的离子化程度也有影响。在流动相中添加无机盐，如硫酸钠，会改变其选择性和保留时间。流动相中的盐及离子对试剂的浓度对柱填料寿命有一定的影响，因此使用后应先用水冲洗较长时间。

本实验采用以1-正辛烷磺酸为离子对试剂的反相高效液相色谱方法来分析2-氨基-1,3,4噻二唑含量。

三、仪器与试剂

1. 仪器

安捷伦1200 HPLC高效液相色谱仪，VWD紫外-可见光检测器，手动进样器，SB3200超声波清洗器。

2. 试剂

2-氨基-1,3,4噻二唑标准品，1-正辛烷磺酸（色谱纯），甲醇（色谱纯），超纯水。

用甲醇配制10μg/mL的2-氨基-1,3,4噻二唑标准母液，放入冰箱中储存，使用时用甲醇稀释配制成一定浓度的2-氨基-1,3,4噻二唑标准溶液。

四、实验步骤

1. 色谱条件

Partisil ODS-3分析柱：5μm，250mm×4.6mm，流动相为含有0.005mol/L 1-正辛烷磺酸的水-甲醇（78：22，体积比），pH=3.40，流速1mL/min，紫外检测波长254nm，0.02 AUFS进行检测；柱温为室温；进样量为20μL。

2. 建立标准工作曲线

以流动相配制不同浓度（50μg/mL、100μg/mL、150μg/mL、200μg/mL、250μg/mL）的敌枯唑标准溶液注入高效液相色谱仪，用上述液相条件测得相应峰面积，建立标准工作曲线。

3. 样品分析

取重结晶后的2-氨基-1,3,4噻二唑合成产物，以流动相配制200μg/mL的样品溶液，进样分析。计算样品相对纯度（相对于实验对照品）。

五、实验HSE

1. 2-氨基-1,3,4噻二唑：有毒物品。毒性分级：高毒。急性毒性，腹腔-大鼠LD50：200mg/kg；腹腔-小鼠LD50：6500mg/kg。可燃性危险特性：可燃；燃烧产生有毒氮氧化物和硫氧化物烟雾。储运特性：库房通风低温干燥。灭火剂：干粉、泡沫、砂土、二氧化碳、雾状水。造成皮肤刺激，造成严重眼刺激，可引起呼吸道刺激。急救措施：皮肤接触者立即脱去污染的衣着，用大量流动清水冲洗至少15min；眼睛接触时立即提起眼睑，用大量流动清水或生理盐水彻底冲洗至少15min；吸入则应迅速脱离现场至空气新鲜处，保持呼吸道通畅，如呼吸困难给输氧，如呼吸停止立即进行人工呼吸；食入应立即用水漱口，严重时就医。储存于阴凉、通风的库房。应与强氧化剂、食用化学品等分开存放。保持容器密封，远离火种、热源。防范说明：作业后彻底清洗。使用本产品时不要进食、饮水或吸烟。戴防护手套、穿防护服、戴防护眼罩、戴防护面具。避免吸入粉尘、烟、气体、烟雾、蒸气、喷雾。只能在室外或通风良好处使用，存放处须加锁。

2. 1-正辛烷磺酸：可引起呼吸道刺激。不可燃。在火场中可产生有害气体或蒸气。储存：干燥，密封，储存温度为15～25℃。急救措施：皮肤接触者立即脱去污染的衣着，用大量流动清水冲洗至少15min；眼睛接触时立即提起眼睑，用大量流动清水或生理盐水彻底冲洗至少15min；吸入则应迅速脱离现场至空气新鲜处，保持呼吸道通畅，如呼吸困难给输氧，如呼吸停止立即进行人工呼吸；如遇食入应立即用水漱口，严重时就医。储存于阴凉、通风的库房。应与强氧化剂、食用化学品等分开存放。保持容器密封，远离火种、热源。

3. 甲醇：有毒、易燃，其蒸气与空气能形成爆炸混合物。甲醇的健康危害主要是对中枢神经系统有麻醉作用；对视神经和视网膜有特殊选择作用，引起病变；可致代谢性酸中毒。人误饮5～10mL甲醇就会严重中毒，造成双目失明，大量饮入会导致死亡。空气中常规允许浓度为5mg/m^3；甲醇工业废水经处理才能排放，处理后的甲醇允许含量不应大于200mg/L。在有甲醇蒸气的现场短时间工作时，要佩戴防毒面具、橡皮手套、防护眼镜等安全用具。急救措施：皮肤接触者立即脱去污染的衣着，用大量流动清水冲洗至少15min；眼睛接触时立即提起眼睑，用大量流动清水或生理盐水彻底冲洗至少15min；吸入则应迅速脱离现场至空气新鲜处，保持呼吸道通畅，如呼吸困难给输氧，如呼吸停止立即进行人工呼吸；食入应立即用水漱口，严重时就医。储存于阴凉、通风的库房。应与强氧化剂、食用化学品等分开存放。保持容器密封，远离火种、热源。

六、思考题

1. 离子对色谱法的适用范围是什么？
2. 相比于离子交换色谱，离子对色谱的优点是什么？
3. 根据离子对色谱分离原理解释为什么要引入离子对萃取原理到反相键合相色谱中。
4. 影响离子对色谱分离选择性的因素有哪些？

七、注意事项

由于反相离子色谱法中使用了离子对试剂和无机盐，因此在操作上要注意以下几个方面：

（1）改变淋洗液浓度或梯度洗脱操作后，柱系统的冲洗时间要长些，以保证柱内平衡；

（2）实验停止后须用纯水或甲醇/水冲洗进样阀、高压泵及色谱柱，以防流动相腐蚀设备和担体；

（3）一经发现离子对试剂在柱内产生了不可逆吸附，则必须用甲醇冲洗柱子；

（4）由于离子对试剂不容易从色谱柱中完全洗脱出来，因此建议柱子应尽可能专用；

（5）含有离子对试剂的流动相，不能直接切换，要先用含水比例高的流动相过渡，然后再切换到含离子对试剂的流动相，一般总平衡时间比单溶剂水延长100%～150%；

（6）在不损害色谱柱稳定性的前提下，淋洗液的pH值应使被分离物质达到最大解离度。为了避免盐析效应，缓冲液的盐浓度应在5×10^{-3}～1×10^{-2}mol/L范围内。

Experiment 10 Ion pair reversed phaser chromatographic separation and quantitation of 2-amino-1,3,4-thiadiazole

1. Objectives

1.1 Master the principle and application of pair reversed phaser chromatographic separation.

1.2 Consolidate the operational skills of HPLC instrument and the notices.

2. Principle

2-amino-1,3,4-thiadiazole have a good effect on the control of rice bacterial blight and other crop bacterial diseases. In the past few years, it has been widely used in China. But animal experiments have shown that 2-amino-1, 3, 4-thiadiazole has a strong teratogenic effect. Some people exposed to these compounds have been found varying degrees of glossitis, cuticular dermatitis, hyperuricemia, and erythema that appeared on the face and genitals. It is necessary to analyze the residue in environmental monitoring. Its polarity is strong and has certain alkalinity. In the silica gel column using chloroform, isopropanol acetic acid (40 : 10 : 1; $V/V/V$) as the mobile phase, the peaks show the asymmetry, the tailing phenomenon is more serious. On the reverse ODS column, methanol-water or acetonitrile-water used as the mobile phase with the system is not retained. Xue et al. used trichloromethane-ethanol-dimethylformamide (95 : 5 : 0.2; $V/V/V$) as the mobile phase, amino bonded phase as the stationary phase for separation, analysis of 2-amino-1,3,4-thiadiazole and its metabolites. Ackerly et al. added octane sulfonate as an ion-pair reagent. The 2-amino-1,3,4-thiadiazole in plasma can be separated and determined on the C18 column. Huang et al. also used the ODS column, which contains 0.005mol/L ion pair reagent of *n*-octane sulfonic acid in methanol as the mobile phase for the detection of 2-amino-1,3,4-thiadiazole.

When the polarity of the stationary phase is less than that of the mobile phase, the separation mode of liquid chromatography is called reversed-phase chromatography. After adding ion pair reagent in the mobile phase, it can combine with the analyzed substance to form ionic pair and become neutral, so that the non-polarity appears, thus increasing the retention on the reversed chromatographic column. This separation pattern is called reversed-phase ion-pair chromatography, and there are two main types of ion-pair reagents, tetrabutylamine phosphate, and alkyl sulfonate. Tetramine salts at pH = 7.5 can form ion pairs with strong acid or weak acid and have the function of buffering and inhibiting weak base. Alkyl sulfonates buffer weak acid ions with a strong base or weak base at pH=3.5. In the reversed-phase ion-pair chromatography, three factors affect the retention and separation of the mobile phase for solutes, namely the composition of the mobile phase, the concentration of counterion, and the organic modifier. The most commonly used mobile phase of the reversed-phase ion-pair chromatography is the buffer solution with water as the main body, and its pH value has the greatest influence on the separation because the pH value of the mobile phase determines the dissociation state of the analyte which affects the degree of its binding to the ions. Therefore, changing the pH value of the mobile phase is an effective way to improve the separation selectivity, the concentration of counterion is the main factor affecting the retention. Increasing the concentration of counterion can increase the retention effect and the polar. The organic solvent can be added to the mobile phase with water as the main body and methanol, acetonitrile as a modifier. Its role is to improve the association of ions in the two phases and have an impact on the degree of ionization of some compounds. Adding inorganic salt (eg. sodium sulfate) to the mobile phase would change its selectivity and retention time. The concentration of salt and counterion in the mobile phase has a certain effect on the life of the column filler, so it should be washed with water for a long time.

Here, we will use a reversed-phase high-performance liquid chromatography method with *n*-octane sulfonic acid as an ion-pair reagent to determine the content of 2-amino-1,3,4-thiadiazole.

3. Instruments and reagents

3.1 Instruments

Agilent 1200 HPLC, VWD UV-Vis detector, manual sampler, SB3200 ultrasonic cleaner.

3.2 Reagents

2-amino-1,3,4-thiadiazole standard, *n*-octane sulfonic acid, methanol (chromatographic purity).

10μg/mL of 2-amino-1,3,4-thiadiazole mother standard was prepared with methanol and stored in a refrigerator. When used, the standard mother liquor of a certain concentration is prepared by diluting methanol.

4. Procedures

4.1 Chromatographic conditions

The analysis column is Partisil 5 microns ODS-3 250mm×4.6mm. The mobile phase is methanol in water (78 : 22, *V*/*V*) containing 0.005mol/L of *n*-octane sulfonic acid. pH= 3.40, flow rate: 1mL/min, UV detection wavelength: 254nm, 0.02AUFS. Room temperature; Sample quantity: 20μL.

4.2 Establishing of standard working curve

A series of concentration evaluation standards (50μg/mL、100μg/mL、150μg/mL、200μg/mL、250μg/mL in mobile phase) to be injected in HPLC, using the above liquid conditions to determine the corresponding peak area to establish a standard working curve.

4.3 Samples analysis

The sample products after recrystallization were analyzed in agriculture with 200μg/mL of the mobile phase. Calculate the relative purity of the sample (with the experimental control).

5. Experimental HSE

5.1 2-amino-1,3,4 thiadiazole: toxic substances. Toxicity grade: high toxicity. Acute toxicity: intraperitoneal-rat LD50: 200mg/kg; Abdominal cavity-mouse LD50: 6500mg/kg. Flammability Hazard Characteristics: Combustibility; Combustion produces toxic nitrogen oxides and sulfur oxide smoke. Storage and transportation characteristics: the warehouse is ventilated and dry at low temperatures. Extinguishing agent: dry powder, foam, sandy soil, carbon dioxide, and atomized water. Summary of Emergency: Harmful Swallowing. Cause skin irritation. Causing severe eye irritation. Cause respiratory tract irritation. First-aid measures: skin contact, immediately takes off contaminated clothes and flush the skin with a large amount of flowing clean water for at least 15 minutes. Lift eyelids immediately when eyes touch, and flush thoroughly with plenty of flowing clear water or normal saline for at least 15 minutes. Inhalation, should be quickly separated from the site to fresh air to keep the respiratory tract unobstructed. If breathing is difficult, give oxygen. If breathing stops, carry out artificial respiration immediately. In the case of food intake, rinse the mouth with water immediately. In case of serious illness, seek medical treatment. Store in a cool and ventilated warehouse. It should be stored separately from strong oxidants, edible chemicals, etc. Keep the container sealed and away from kindling and heat sources. Precautionary instructions: Clean thoroughly after an operation. Do not eat, drink, or smoke when using this product. Wear protective gloves, clothing, goggle, and masks. Avoid inhaling dust, smoke, gas, smoke, vapor, and spray. Can only be used outdoors or in well-ventilated places. Store in a well-ventilated place. Keep the container closed. The storage must be locked.

5.2 1-Octanesulfonic Acid: May cause respiratory tract irritation. Not flammable. Harmful gases or vapors may be generated in the fire scene. Storage: dry and sealup. The storage temperature is between 15℃ and 25℃. First-aid measures: skin contact immediately takes off contaminated clothes and flush with a large amount of flowing clean water for at least 15 minutes. Lift eyelids immediately when eyes touch, and flush thoroughly with plenty of flowing clear water or normal saline for at least 15 minutes. Inhalation, should be quickly separated from the scene to go out for some fresh air to keep the respiratory tract unobstructed. If breathing is difficult, give oxygen. If breathing stops, carry out artificial respiration immediately. In the case of food intake, flush the mouth with water immediately. In case of serious illness, seek medical treatment. Store in a cool and ventilated warehouse. It should be stored separately from strong oxidants, edible chemicals, etc. Keep the container sealed and away from kindling and heat sources.

5.3 Methanol: toxic, flammable, its steam and air can form an explosive mixture. The health hazards of methanol are mainly anesthetic effects on the central nervous system. It has a special selective effect on the optic nerve and retina, causing pathological changes. Can cause metabolic acidosis. People who drink 5～10mL methanol by mistake will be seriously poisoned resulting in blindness, and death due to a large amount of drinking. The conventional allowable concentration in the air is $5mg/m^3$; Methanol industrial wastewater can only be discharged after treatment. The allowable content of methanol after treatment should not be greater than 200mg/L. Wear gas masks, rubber gloves, protective glasses, and other safety appliances when working for a short time in the field with methanol vapor. Packaging containers shall be marked with "flammable liquid", "drugs" and other dangerous goods. First-aid measures: skin contact, immediately takes off contaminated clothes and flush with a large amount of flowing clean water for at least 15 minutes. Lift eyelids immediately when eyes touch, and flush thoroughly with plenty of flowing clear water or normal saline for at least 15 minutes. Inhalation, should be quickly separated from the scene to go out for some fresh air to keep the respiratory tract unobstructed. If breathing is difficult, give oxygen. If breathing stops, carry out artificial respiration immediately. In the case of food intake, rinse the mouth with water immediately. In case of serious illness, seek medical treatment. Store in a cool and ventilated warehouse. It should be stored separately from strong oxidants, edible chemicals, etc. Keep the container sealed and away from kindling and heat sources.

6. Questions

6.1 What is the scope of application of ion-pair chromatography?

6.2 What are the advantages of ion-pair chromatography over ion-exchange chromatography?

6.3 According to the principle of ion-pair chromatography, explain why the principle of ion-pair extraction should be introduced into reversed-phase bonded phase chromatography.

6.4 What are the factors that affect the selectivity of ion-pair chromatography?

7. Notes

Due to the use of ion-pair reagents and inorganic salts in the RP-HPLC, attention should be paid to the following aspects in the operation.

7.1 After changing the operation of flushing purulent fluid or gradient flushing, the washing time of the system should be longer to ensure the balance in the column.

7.2 After the experiment is stopped, use pure water or methanol-water, flush sample valve, high-pressure pump, and column to prevent the mobile phase from corroding equipment and monomer.

7.3 Once the irreversibly adsorbed ion pair reagent is found in the column, the column should be washed with methanol.

7.4 Since it is not easy for you to completely clean the reagent from the chromatographic column, it is recommended that the column using the reagent with ion-pair should be as exclusive as possible.

7.5 The mobile phase containing ion pair reagent can not be directly switched, so the mobile phase containing ion pair reagent. Generally, the total equilibrium time is 100%～150% longer than that of only water as a solution.

7.6 The pH value of the buffer solution of the flushing agent should reach the maximum dissociation without damaging the stability of the chromatographic column. To avoid the salting-out effect, the salt concentration of the buffer should be within the range of 1×10^{-2} mol/L to 5×10^{-3} mol/L.

实验十一　聚丙烯酸合成

一、实验目的

1. 掌握低分子量聚丙烯酸合成。

2. 掌握自由基聚合原理、基本的溶液聚合方法。

二、实验原理

聚丙烯酸是水质稳定剂的主要原料之一。高分子量的聚丙烯酸（分子量在几万或几十万以上）多用于皮革工业、造纸工业等。作为阻垢用的聚丙烯酸，分子量都在一万以下。聚丙烯酸分子量的大小对阻垢效果有极大影响。低分子量聚丙烯酸阻垢作用显著，而高分子量聚丙烯酸丧失阻垢作用。

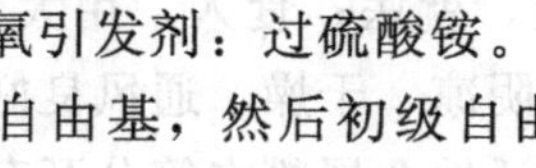

聚合方法：丙烯酸单体聚合，可以通过本体、溶液、乳液和悬浮等聚合方法得到聚丙烯酸。符合一般的自由基聚合规律。本实验采用溶液聚合。

反应原理：本实验采用无机过氧引发剂：过硫酸铵。

首先是过硫酸铵分解形成初级自由基，然后初级自由基和丙烯酸单体加成形成单体自由基。

引发剂分解是吸热反应，活化能高，初级自由基与单体结合这一步是放热反应，自由基活性高，有相互作用引起终止的倾向。终止反应有耦合终止和歧化终止两种方式。链终止的方式与单体种类和聚合条件有关。中间可能会发生链转移。

反应影响因素：本实验的目的是得到低分子量的聚丙烯酸。而控制聚合度和聚合速率，需要从引发剂浓度和聚合温度两个因素综合考虑。

链转移（可能向引发剂转移，使引发剂的效率降低）的结果，很可能产生阻垢作用，使分子量较低。提高温度一般可以使链转移常数增加。凝胶效应使分子量增加。

三、仪器与试剂

1. 仪器

四口瓶，回流冷凝管，电动搅拌器，恒温水浴锅，滴液漏斗。

2. 试剂

丙烯酸，过硫酸铵，异丙醇。

四、实验步骤

1. 装有搅拌器、回流冷凝管、滴液漏斗和温度计的250mL四口瓶中，加入50mL蒸馏水和0.5g过硫酸铵。待过硫酸铵溶解后，加入3.0g丙烯酸和4.0g异丙醇。开动电动搅拌器，加热使反应瓶内温度控制在65～70℃。

2. 将20g丙烯酸单体和2.0g过硫酸铵在40mL水中溶解，用滴液漏斗逐渐滴入四口瓶内。由于聚合过程放热，瓶内温度有所升高，反应逐渐回流。滴完丙烯酸单体和过硫酸铵溶液大约需要0.5h。

3. 在90℃回流1h，反应完成。(聚丙烯酸酯分子量为500～4000)。

五、实验 HSE

1. 丙烯酸：无色液体。强有机酸。有刺激性酸味，产生烟雾。急性毒性：烟雾对眼睛、鼻黏膜有刺激性，沾在皮肤上对皮肤有腐蚀性及刺激性，操作时应注意。大鼠经口 LD50：2590mg/kg。急救措施：皮肤接触，立即脱去污染的衣着，用流动清水彻底冲洗皮肤至少15min，就医；眼睛接触，立即提起眼睑，用大量流动清水或生理盐水彻底冲洗至少15min，就医；吸入，迅速脱离现场至空气新鲜处，保持呼吸道通畅，如呼吸困难给输氧，如呼吸停止立即进行人工呼吸，就医；食入，饮足量温水，催吐，就医。储存注意事项：有腐蚀性，化学性质活泼，不稳定，遇氧易聚合，保存时可加入氢醌或甲氧基氢醌等阻聚剂。

2. 过硫酸铵：健康危害：对皮肤黏膜有刺激性和腐蚀性。吸入后引起鼻炎、喉炎、气短和咳嗽等。眼、皮肤接触可引起强烈刺激、疼痛甚至灼伤。口服引起腹痛、恶心和呕吐。长期皮肤接触可引起反应性皮炎。急性毒性：LD50：820mg/kg（大鼠经口）。急救措施：皮肤接触，立即脱去污染的衣着，用大量流动清水冲洗至少15min，可涂抹弱碱性物质（如碱水、肥皂水等），就医；眼睛接触，立即提起眼睑，用大量流动清水或生理盐水彻底冲洗至少15min，就医；吸入，迅速脱离现场至空气新鲜处，保持呼吸道通畅，如呼吸困难给输氧，如呼吸停止立即进行人工呼吸，就医；食入，用大量水漱口，吞服大量生鸡蛋清或牛奶，就医。储存注意事项：储存于阴凉、干燥、通风良好的库房。远离火种、热源。包装必须密封，防止受潮。应与还原剂、活性金属粉末等分开存放，切忌混储。储区应备有合适的材料收容泄漏物。

3. 异丙醇：有似乙醇和丙酮混合物的气味。急性毒性：微毒性，LD50：5840mg/kg（大鼠经口）。高浓度蒸气具有明显麻醉作用，对眼，呼吸道的黏膜有刺激作用，能损伤视网膜及视神经。急救措施：皮肤接触，立即脱去污染的衣着，用大量流动清水冲洗，就医；眼睛接触，立即提起眼睑，用大量流动清水或生理盐水彻底冲洗，就医；吸入，迅速脱离现场至空气新鲜处，保持呼吸道通畅，如呼吸困难给输氧，如呼吸停止立即进行人工呼吸，就医；食入，饮用大量清水，催吐，就医洗胃。储存注意事项：该品应密封于阴凉干燥处避光生存。远离热源。储存温度不超过 30℃。可用铁、软钢、铜或铝制容器储存。

4. 注意丙烯酸单体和过硫酸铵混合溶液滴加速率，滴加过程中可能会发生恒压滴液漏斗堵塞，注意实验观察。全程不得离开实验室。

六、思考题

1. 本实验采用的聚合方法是什么？
2. 如何控制聚丙烯酸的低分子量？
3. 除了本实验中的过硫酸铵，还可以使用哪些引发剂？

Experiment 11 Synthesis of Polyacrylic acid

1. Objectives

1.1 Preparation of low molecular weight polyacrylic acid.

1.2 Master the principle of free radical polymerization and a basic solution polymerization method.

2. Principle

Polyacrylic acid is one of the main raw materials for water stabilizers. High molecular weight polyacrylic acid (relative molecular weight of tens of thousands or hundreds of thousands) is used in the leather industry, paper industry, and the like. As the polyacrylic acid for scale inhibition, the relative molecular weight is 10,000 or less. The relative molecular weight of polyacrylic acid has a great influence on the scale inhibition effect. Low molecular weight polyacrylic acid has a significant scale inhibition effect, while high molecular weight polyacrylic acid loses its scale inhibition effect.

In this experiment, low molecular weight polyacrylic acid was synthesized by controlling the amount of initiator and using the telogen isopropanol.

Aggregation method: The acrylic monomer is polymerized, and polyacrylic acid can be obtained by a polymerization method such as bulk, solution, emulsion, and suspension. It conforms to the general rule of free radical polymerization. This experiment uses solution polymerization.

Reaction principle: This experiment uses an inorganic peroxygenic initiator: ammonium persulfate.

The first procedure is the decomposition of the primary free radicals by ammonium persulfate. The primary free radical and the acrylic monomer are then added to form a monomeric radical.

The decomposition of the initiator is an endothermic reaction, and the activation energy is high. The step of combining the primary radical with the monomer is an exothermic reaction, and the radical activity is high, and there is a tendency for the interaction to terminate. The termination reaction has two modes: coupling termination and disproportionation termination. The manner in which the chain terminates is related to the type of monomer and the polymerization conditions. Chain transfer may occur in the middle.

Reaction factors: The purpose of this experiment is to obtain a low molecular weight polyacrylic acid. Controlling the degree of polymerization and the rate of polymerization requires a combination of both initiator concentration and polymerization temperature.

As a result of chain transfer (possible transfer to the initiator, which reduces the efficiency of the initiator), it is likely to cause scale inhibition and lower molecular weight. Increasing the temperature generally increases the chain transfer constant. The gel effect increases the molecular weight.

3. Instruments and Reagents

3.1 Instruments

Four-port flask, reflux condenser, electric mixer, constant temperature water bath, dropping funnel.

3.2 Reagents

Acrylic acid, ammonium persulfate, isopropanol.

4. Procedures

4.1 Add 250mL of distilled water and 0.5g of ammonium persulfate to a 250mL four-necked flask equipped with a stirrer, reflux condenser, dropping funnel, and thermometer. After the ammonium persulfate was dissolved, 3.0g of acrylic acid and 4.0g of isopropanol were added. Start the electric agitator and heat to keep the temperature inside the reaction flask at 65～70℃.

4.2 Dissolve 20g of acrylic acid monomer and 2.0g of ammonium persulfate in 40mL of water, and gradually drip into a four-port flask with a dropping funnel. Due to the exothermic heat of the polymerization process, the temperature inside the flask increased and the reaction gradually returned. It takes about 0.5h to complete the acrylate monomer and ammonium persulfate solution.

4.3 Reflow at 90℃ for 1 hour and then the reaction is complete. (The relative molecular weight of polyacrylate is 500～4000).

5. Experimental HSE

5.1 acrylic acid: Corrosive, chemically active, unstable, easy to polymerize in the presence of oxygen. Acute toxicity: smoke is irritant to the eyes and nasal mucosa, and it is corrosive and irritant to the skin when touching skin. Attention should be paid during operation. LD50=2590mg/kg. **First-aid measures**: skin contact, take off the contaminated clothes immediately, wash the skin thoroughly with flowing water and clear water for at least 15 minutes, and get medical attention. Eye contact, lift the eyelid immediately, wash thoroughly with a large amount of flowing water or normal saline for at least 15 minutes, and seek medical treatment. Inhalation, quickly leave the scene to go out for some fresh air. Keep the respiratory tract unobstructed. If breathing is difficult, give oxygen. If breathing stops, perform artificial respiration immediately. Get medical attention. Ingestion: drink enough warm water, induce vomiting, and seek medical treatment. **Storage precautions**: corrosive, chemical active, unstable, easy to polymerize in the presence of oxygen, adding hydroquinone or methoxyhydroquinone, and another polymerization inhibitor during storage.

5. 2 ammonium persulfate: Health hazard: irritating and corrosive to the skin and mucous membrane. Cause rhinitis, laryngitis, shortness of breath, and cough after inhalation. Eye and skin contact can cause strong irritation, pain, and even burns. Oral administration causes abdominal pain, nausea, and vomiting. Long term skin contact can cause allergic dermatitis.

Acute toxicity: LD50=820mg/kg (rat oral) . **First-aid measures**: skin contact, take off the contaminated clothes immediately, wash with a large amount of flowing water for at least 15 minutes, smear weak alkaline substances (such as alkali water, soap water, etc.) and seek medical treatment. Eye contact, lift the eyelid immediately and wash thoroughly with a large amount of flowing water or normal saline for at least 15 minutes. Get medical attention. Inhalation, quickly leave the site to fresh air. Keep the respiratory tract unobstructed. If breathing is difficult, give oxygen. If breathing stops, perform artificial respiration immediately. Get medical attention. Ingestion: gargle with plenty of water, swallow a lot of raw egg white or milk, and seek medical treatment. **Storage precautions**: Store in a cool, dry, and well-ventilated warehouse. Keep away from fire and heat. Packaging must be sealed to prevent moisture. It should be stored separately from reducing agent and active metal powder, and avoid mixed storage. The storage area shall be equipped with appropriate materials to contain the leakage.

5. 3 isopropanol: Acute toxicity: slight toxicity, LD50: 5840mg/kg (rat oral). High concentration steam has an obvious anesthetic effect, irritating effect on the mucous membrane of the eyes and respiratory tract, and can damage the retina and optic nerve. **First-aid measures**: skin contact, immediately remove the contaminated clothes, wash with a large amount of flowing water, and seek medical treatment. Eye contact, lift the eyelid immediately and wash thoroughly with a large amount of flowing water or physiological saline. Get medical attention. Inhalation, quickly leave the scene to go out for some fresh air. Keep the respiratory tract unobstructed. If breathing is difficult, give oxygen. If breathing stops, perform artificial respiration immediately. Get medical attention. Ingestion: drink plenty of clear water, induce vomiting, seek medical treatment, and gastric lavage. **Storage precautions**: the product should be sealed in a cool and dry place away from light. Keep away from heat. The controlled temperature shall not exceed 30℃. It can be stored in iron, mild steel, copper, or aluminum containers.

5. 4 Pay attention to the drop acceleration rate of the mixed solution of acrylic monomer and ammonium persulfate, which may cause constant pressure to drop funnel to be blocked during the dropping process. Pay attention to the experimental observation. Do not leave the laboratory during the whole process.

6. Questions

6.1 What is the polymerization method used in this experiment?

6.2 How to control the low molecular weight of polyacrylic acid?

6.3 In addition to ammonium persulfate in this experiment, what other initiators can be used?

实验十二　黏度法测定水溶性高聚物的平均分子量

一、实验目的

1. 掌握黏度法测定高聚物平均分子量的基本原理。
2. 掌握运用乌氏黏度计测定黏度的实验技术及数据处理方法。

二、实验原理

高聚物的研究中，分子量是一个不可缺少的重要数据。因为它不仅反映了高聚物分子的大小，并且直接关系到高聚物的物理性能。但与一般的无机物或低分子有机物不同，高聚物多是分子量不等的混合物，因此通常测得的分子量是一个平均值。高聚物分子量的测定方法很多，比较起来，黏度法设备简单，操作方便，并有很好的实验精度，是常用的方法之一。

高聚物稀溶液的黏度是它在流动过程所存在的内摩擦的总体反映。这种流动过程中的内摩擦主要有：溶剂分子之间的内摩擦；高聚物分子与溶剂分子间的内摩擦；高聚物分子之间的内摩擦。其中溶剂分子之间的内摩擦又称为纯溶剂的黏度，以 η_0 表示。三种内摩擦的总和称为高聚物溶液的黏度，以 η_s 表示，实践证明，在同一温度下，高聚物溶液的黏度一般要比纯溶剂的黏度大些，即 $\eta > \eta_0$。为了比较这两种黏度，引入增比黏度的概念，以 η_{sp} 表示。

$$\eta_{sp}=\frac{\eta-\eta_0}{\eta_0}=\frac{\eta}{\eta_0}-1=\eta_r-1 \tag{1}$$

式中，η_r 为相对黏度，它是溶液黏度与溶剂黏度的比值，反映的仍是整个溶液黏度的行为；η_{sp}则反映扣除了溶剂分子间的内摩擦后仅仅是纯溶剂与高聚物分子间以及高聚物分子之间的内摩擦。显而易见，高聚物溶液的浓度变化，将会直接影响 η_{sp} 的大小，浓度越大，黏度也越大。为此，常常取单位浓度下呈现的黏度来进行比较，从而引入比浓黏度的概念，以$\frac{\eta_{sp}}{c}$表示。又将$\frac{\ln\eta_{sp}}{c}$定义为比浓对数黏度。因为 η_r 和 η_{sp} 是无因次量，$\frac{\eta_{sp}}{c}$和$\frac{\ln\eta_{sp}}{c}$的单位由浓度的单位而定，通常采用 g/mL。为了进一步消除高聚物分子间内摩擦的作用，必须将溶液无限稀释，当浓度 c 趋近于零时，比浓黏度趋近于一个极限值，即：

$$\lim \frac{\eta_{sp}}{c}=[\eta] \tag{2}$$

$[\eta]$ 主要反映了高聚物分子与溶剂分子之间的内摩擦作用，称之为高聚物溶液的特性黏度，其数值可通过实验求得。因为根据实验，在足够稀的溶液中有：

$$\frac{\eta_{sp}}{c}=[\eta]+K[\eta]^2c \tag{3}$$

$$\frac{\ln\eta_r}{c}=[\eta]-\beta[\eta]^2c \tag{4}$$

这样以$\frac{\eta_{sp}}{c}$及$\frac{\ln\eta_r}{c}$对c作图得两条直线，这两条直线在纵坐标上相交于同一点（如图1所示），可求出［η］数值。为了绘图方便，引进相对浓度c'，即$c'=c/c_1$。其中，c表示溶液的真实浓度，c_1表示溶液的起始浓度。

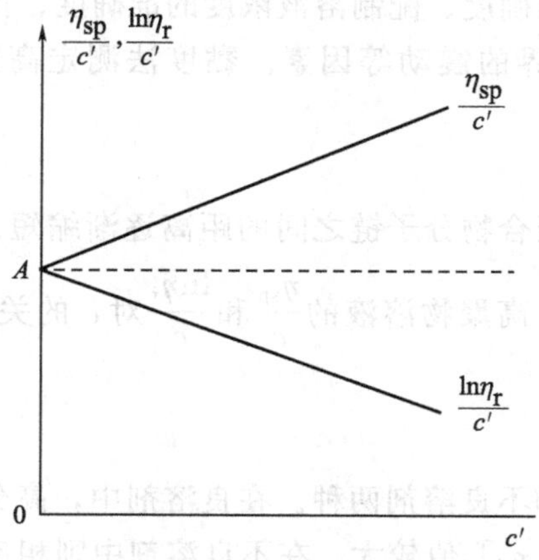

图1　增比黏度和相对黏度与c的关系

由图可知，

$$[\eta]=\frac{A}{c_1} \tag{5}$$

其中A为截距。

由溶液的特性黏度［η］还无法直接获得高聚物分子量的数据，目前常用的由半经验的麦克（H. Mark）非线性方程来求得：

$$[\eta]=KM^{\alpha} \tag{6}$$

式中，M为高聚物分子量的平均值；K、α为常数，与温度、高聚物性质、溶剂等因素有关，可通过其他方法求得。实验证明，α值一般在0.5～1之间。聚乙烯醇的水溶液在25℃时，$\alpha=0.76$，$K=2\times10^{-2}$；在30℃时，$\alpha=0.64$，$K=6.66\times10^{-2}$。式(6)适用于非支化的、聚合度不太低的高聚物。

由上述可以看出高聚物分子量的测定最后归结为溶液特征黏度［η］的测定。而黏度的测定可以按照液体流经毛细管的速度来进行，根据泊塞勒（poiseuille）公式

$$\eta=\frac{\pi r^4 thg\rho}{8lV} \tag{7}$$

式中，V为流经毛细管液体的体积；r为毛细管半径；ρ为液体密度；l为毛细管的长度；t为流出时间；h是作用于毛细管中溶液上的平均液柱高度，$h=(h_1+h_2)/2$；g为重力加速度。

液体在毛细管内靠液柱的重力流动，它所具有的位能，除了消耗于克服分子内摩擦的阻力外，同时使液体本身获得了动能，使实际测得的液体黏度偏低。液体的流速较大时，动能

消耗的能量可达 20%。因此，对泊塞勒公式必须进行修正。液体流动较慢时，动能消耗很小，可以忽略。这时，对于同一黏度计来说 h、r、l、v 是常数，则式(7) 可简化为

$$\eta = k'\rho t \tag{8}$$

考虑到通常测定是在高聚物的稀溶液下进行的，溶液的密度 ρ 与纯溶剂的密度 ρ_0 可视为相等，则溶液的相对黏度就可表示为：

$$\eta_r = \frac{\eta_s}{\eta_0} = \frac{k'\rho t}{k'\rho_0 t_0} \approx \frac{t}{t_0} \tag{9}$$

由此可见，由黏度法测高聚物分子量，最基础的测定是 t_0、t、c，实验的成败和准确度取决于测量液体所流经的时间的准确度、配制溶液浓度的准确度、恒温槽的恒温程度、安装黏度计的垂直位置的程度以及外界的震动等因素。黏度法测定高聚物分子量时，要注意以下几点。

（1）溶液浓度的选择

随着溶液浓度的增加，聚合物分子链之间的距离逐渐缩短。因而分子链间作用力增大。当溶液浓度超过一定限度时，高聚物溶液的 $\frac{\eta_{sp}}{c}$ 和 $\frac{\ln\eta_r}{c}$ 对 c 的关系不呈线性。通常选用 $\eta_r = 1.2 \sim 2.0$ 的浓度范围。

（2）溶剂的选择

高聚物的溶剂有良溶剂和不良溶剂两种。在良溶剂中，高分子线团伸展，链的末端距增大，链段密度减少，溶液的 $[\eta_r]$ 值较大。在不良溶剂中则相反，且溶解很困难。在选择溶剂时，要注意考虑溶解度、价格、来源、沸点、毒性、分解性和回收等方面的因素。

（3）毛细管黏度计的选择

常用毛细管黏度计有乌氏和奥氏两种，测分子量选用乌氏黏度计（图 2）。对球体积为 5mL 的黏度计，一般要求溶剂流经时间为 $t = 100 \sim 130$s 之间。

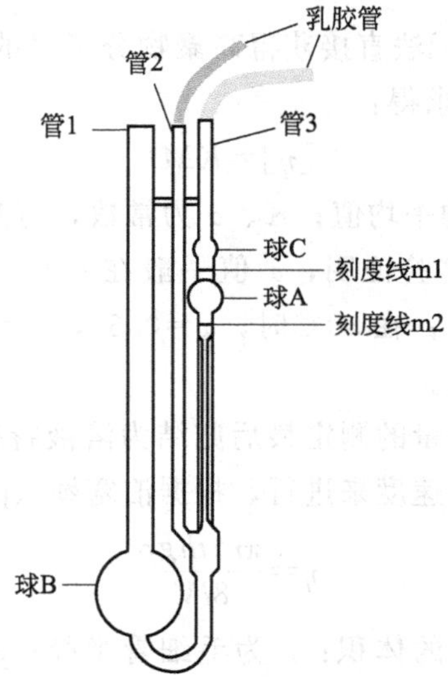

图 2　毛细管黏度计

（4）恒温槽

温度波动直接影响溶液黏度的测定，国家规定用黏度计测定分子量的恒温槽的温度波动

为±0.05℃。

三、仪器与试剂

1. 仪器

恒温槽，乌氏黏度计，停表，吸耳球，容量瓶（100mL），移液管（10mL），烧杯（100mL），玻璃砂漏斗（3号）。

2. 试剂

聚丙烯酸。

四、实验步骤

1. 高聚物溶液的配制

称取0.5g聚丙烯酸放入100mL烧杯中，注入约60mL的蒸馏水，稍加热使之溶解。待冷却至室温，加入2滴正丁醇（去泡剂），并移入100mL容量瓶中，加水至刻度线，记其浓度为c。如果溶液中有固体杂质，用3号玻璃漏斗过滤后待用。过滤不能用滤纸，以免纤维混入。

2. 安装黏度计

所用黏度计必须洁净，有时微量的灰尘、油污等会产生局部的堵塞现象，影响溶液在毛细管中的流速，从而导致较大的误差。所以做实验之前，应该彻底洗净，放在烘箱中干燥，然后在侧管2上端套一软胶管，并用夹子夹紧使之不漏气。调节恒温槽温度至25℃。把黏度计垂直放入恒温槽中，使球C完全浸没在水中，放置位置要合适，便于观察液体的流动情况。恒温槽的搅拌马达的搅拌速度应调节合适，不致产生剧烈震动，影响测定的结果。

3. 溶剂流出时间t的测定

用移液管取10mL蒸馏水由管1（见图2）注入黏度计球B中。待恒温后，利用吸耳球由管3处将溶剂经毛细管吸入球A和球C上部（注意：液体不准吸到吸球内），然后打开侧管2胶管上的夹子，使管2与大气相通，移去管3吸耳球，让溶剂依靠重力自由流下。当液面达到刻度线m1时，立刻按停表开始计时，当液面下降到刻度线m2时，再按停表，记录溶剂流经毛细管的时间t_0。重复三次，每次相差不应超过0.2s，取其平均值。如果相差过大，则应检查毛细管有无堵塞现象；查看恒温槽温度是否符合。

4. 溶液流出时间的测定

待t_0测完后，取10mL配制好的聚丙烯酸溶液加入黏度计球B之中，用吸耳球将溶液反复抽吸至球A内几次，使混合均匀。测定$c'=\frac{1}{2}c$的流出时间t_1，然后再依次加入10mL蒸馏水，稀释成浓度为$\frac{1}{3}c$、$\frac{1}{4}c$、$\frac{1}{5}c$的溶液并分别测定流出时间t_2、t_3、t_4（每个数据重复三次，取其平均值）。

实验完毕，黏度计应洗净，然后用洁净的蒸馏水浸泡或倒置使其晾干。为除掉灰尘的影响，所使用的试剂瓶、黏度计应扣在钟罩内，移液管也应用塑料薄膜覆盖（切勿用纤维材料）。

5. 附数据记录格式及数据处理

（1）将实验数据记录于下表中。

		流出时间				η_r	η_{sp}	$\frac{\eta_{sp}}{c}$	$\ln\eta_r$	$\frac{\ln\eta_r}{c'}$
		测定值			平均值					
		1	2	3						
溶剂					$t_0=$					
溶液	$c'=\frac{1}{2}c$				$t_1=$					
	$c'=\frac{1}{3}c$				$t_2=$					
	$c'=\frac{1}{4}c$				$t_3=$					
	$c'=\frac{1}{5}c$				$t_4=$					

(2) 作$\frac{\eta_{sp}}{c'}$-c'图和$\frac{\ln\eta_{sp}}{c'}$-c'图，并外推至$c'=0$，从截距求出 [η] 值。

(3) 由 [η] $=KM^{\alpha}$式求出聚乙烯醇的分子量M。

五、实验 HSE

1. 做好实验的关键是黏度计必须清洁，直至毛细管壁不挂水珠；恒温槽的温度要控制在±0.1℃内；高聚物在溶剂中溶解缓慢，配制溶液时，要确保其完全溶解，并且如果高聚物中有絮状物，则不能将它移入黏度计中。

2. 每加入一次溶剂稀释时都必须混合均匀。

3. 黏度计必须垂直放置，实验过程中不要振动黏度计。

4. 黏度计是贵重玻璃仪器，在洗涤、安放和使用过程中要小心，避免损坏。

六、思考问题

1. 乌氏黏度计中的侧管 2 有什么作用？本实验能否采用奥氏黏度计（即减去 C 管）？

2. 黏度计的毛细管的粗细对实验结果有何影响？

3. 测定黏度时，黏度计必须一要垂直，二要放入恒温槽内，为什么？

4. 测定相对黏度，可以不用同一根黏度计吗？

5. 评价黏度法测定高聚物分子量的优缺点，指出影响准确测定结果的因素。

七、讨论

1. 最常用的毛细管黏度计有两种。一种是三管黏度计，即本实验采用的乌氏黏度计。其特点是溶液的流出时间与加入 F 球中待测液的体积无关，因而可以在黏度计里加入溶剂或溶液改变待测液的浓度。另一种是二管黏度计，即奥氏黏度计。因为液体的流出时间与加入黏度计中的溶液的液面高度有关，因此，测定时标准液和待测液的体积必须相同。考虑到$\frac{mV}{8\pi lt}$可以忽略，需选择适宜的毛细管长度、直径的大小和 E 球的大小，使流出的时间大于 100s，最好在 120s 左右为宜。但毛细管也不宜太细，否则测定时容易堵塞黏度计。黏度计使用完毕，应立即清洗，防止聚合物黏结甚至堵塞毛细管孔径。清洗后在黏度计内注满蒸馏水并加塞，防止落进灰尘。

2. 特性黏度的单位与浓度的单位互为倒数，然而在文献和现发行的实验教材中所用的单位也不完全相同，致使 $[\eta]=K\overline{M}^{\alpha}$ 一式中的常数 K 有数量级的不同。参数 K 的单位与 $[\eta]$ 相同，而参数 α 则是无因次的量。对同一聚合物，在不同的温度和不同的溶剂条件下，K 及 α 的值不同，使用时应注意。

3. 高聚物分子链在溶液中表现出的一些行为会影响 $[\eta]$ 的测定。

（1）聚电解质行为，即某些高分子链的侧基可以电离，电离后的高分子链有相互排斥作用，随 c 的减小，η_{sp}/c 却反常地增大。通常可以加入少量小分子电解质作为抑制剂，利用同离子效应加以抑制。

（2）某些高聚物在溶液中会发生降解，会使 $[\eta]$ 和 M 结果偏低，可加入少量的抗氧剂加以抑制。

4. 以 η_{sp}/c-c 及 η_r/c-c 作图不呈线性的影响因素。

（1）温度的波动

一般而言，对于不同的溶剂和高聚物，温度的波动对黏度的影响不同。溶液黏度与温度的关系可以用 Andraole 方程 $\eta=Ae^{B/(RT)}$ 表示，式中，A 与 B 对于给定的高聚物和溶剂是常数；R 为气体常数。因此，这要求恒温槽具有较好的控温精度。

（2）溶液的浓度

随着浓度的增加，高聚物分子链之间的距离逐渐缩短，因而分子链间作用力增大，当浓度超过一定限度时，高聚物溶液的 η_{sp}/c 或 $\ln\eta_r/c$ 与 c 的关系不呈线性。通常选用 $\eta_r=1.2\sim2.0$ 的浓度范围。

（3）测定过程中因为毛细管垂直情况发生改变以及微粒杂质局部堵塞毛细管而影响流经时间。

5. 黏度测定中异常现象的近似处理。

在特定性黏度测量过程中，有时并非操作不慎，而出现对图外推时，在纵坐标轴上并不相交于一点的异常现象。此时途中应该以 η_{sp}/c-c 的关系为基准来求得高聚物溶液的特性黏度即 $[\eta]$。

Experiment 12 Determination of the average relative molecular weight of water-soluble polymer by viscosity method

1. Objectives

1.1 Master the basic principle of determination of the average relative molecular weight of polymer by viscosity method.

1.2 Master the experimental techniques and data processing method of measuring viscosity by using Ubbelohde's viscometer.

2. Principle

Relative molecular weight is an important and indispensable data in polymer research. Because it not only reflects the size of polymer molecules, but also directly related to the physical properties of the polymer. Unlike ordinary inorganic or low molecular organics, high polymers are often mixtures of different molecular weights, so the measured relative molecular weight is an average value. There are many methods to determine the relative molecular weight of the polymer. Compared with other methods, the viscosity method is one of the commonly used methods because of its simple equipment, convenient operation, and good experimental accuracy.

The viscosity of the dilute polymer solution is an overall reflection of the internal friction in the flow process. The internal friction in this flow process mainly includes the internal friction between solvent molecules, the polymer and the solvent molecules, and between the high polymer molecules. Where the internal friction between the solvent molecules is also known as the viscosity of the pure solvent, expressed as η_0. The sum of the three internal frictions is called the viscosity of the polymer solution, expressed as η_s. It has proved that, at the same temperature, the viscosity of the polymer solution is generally larger than that of the pure solvent, that is, $\eta_s > \eta_0$. To compare these two viscosities, the concept of increasing specific viscosity is introduced, expressed as η_{sp}.

$$\eta_{sp} = \frac{\eta_s - \eta_0}{\eta_0} = \frac{\eta_s}{\eta_0} - 1 = \eta_r - 1 \tag{1}$$

where the η_r is called relative viscosity, which is the ratio of the viscosity of the solution to the solvent, reflecting the behavior of the whole solution viscosity. η_{sp} reflects the internal friction between the pure solvent and the polymer molecules and between the polymer molecules after deducting the internal friction between the solvent molecules. The change in the concentration of the polymer solution will directly affect the η_{sp}. Generally, the higher the concentration, the higher the η_{sp}. For this reason, the viscosity per unit concentration is often used for comparison. Thus, the concept of specific viscosity is introduced, expressed as $\frac{\eta_{sp}}{c}$. $\frac{\ln\eta_{sp}}{c}$ is also defined as logarithmic viscosity with a specific concentration. Since η_r and η_{sp} are dimensionless quantities, the units of $\frac{\eta_{sp}}{c}$ and $\frac{\ln\eta_{sp}}{c}$ are determined by the units of concentration, the unit of g/mL is usually adopted.

To eliminate the friction between polymer molecules, the solution must be diluted indefinitely. When the concentration of c tends to zero, the specific viscosity tends to a limit value, namely:

$$\lim_{c \to 0} \frac{\eta_{sp}}{c} = \lim_{c \to 0} \frac{\ln\eta_r}{c} = [\eta] \tag{2}$$

$[\eta]$ mainly reflects the internal friction between polymer molecules and solvent molecules, which is called the characteristic viscosity of polymer solution. Its value can be obtained by experiment. Because according to the experiment, in a sufficiently dilute solution:

$$\frac{\eta_{sp}}{c} = [\eta] + K[\eta]^2 c \tag{3}$$

$$\frac{\ln\eta_r}{c} = [\eta] - \beta[\eta]^2 c \tag{4}$$

In this way, two straight lines are drawn by drawing $\frac{\eta_{sp}}{c}$ and $\frac{\ln\eta_r}{c}$ to c. These two lines intersect with the same point in the y-coordinate (as shown in the fig. 1), and the value of $[\eta]$ can be obtained. For the convenience of plotting, the relative concentration of c' is introduced, that is, $c'=c/c_1$, where c represents the true concentration of the solution, c_1 represents the initial concentration of the solution,

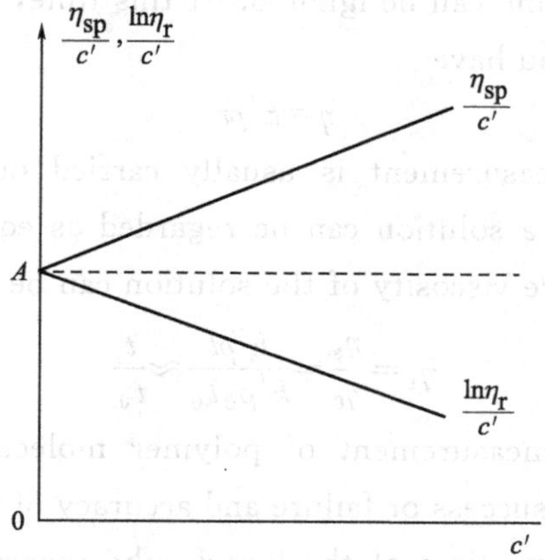

Fig. 1 The relation of the increasing specific viscosity, the relative viscosity and c

According to the figure,

$$[\eta]=\frac{A}{c_1} \tag{5}$$

where A is the intercept. It is not possible to directly obtain the molecular weight data of polymers from the solution's characteristic viscosity $[\eta]$. Currently, it is commonly obtained by semi-empirical H. Mark's nonlinear equation:

$$[\eta]=KM^{\alpha} \tag{6}$$

In the formula, M is the average molecular weight of the polymer, K and α are constants, which are related to temperature, polymer properties, solvent, and other factors, which can be obtained by other methods. Experiments have shown that the values of α are usually between 0.5 and 1. An aqueous solution of polyvinyl alcohol at 25℃, $\alpha=0.76$, $K=2\times10^{-2}$; at 30℃, $\alpha=0.64$, $K=6.66\times10^{-2}$. The formula of (6) applies to non-branched high polymers with a moderate degree of polymerization.

It can be seen from the above that the determination of the polymer's molecular weight can be concluded as the determination of the characteristic viscosity of the solution $[\eta]$. Viscosity can be measured by the velocity at which a liquid flows through the capillary, according to Poiseuille's formula:

$$\eta=\frac{\pi r^4 thg\rho}{8lV} \tag{7}$$

where, V is the volume flowing through the capillary liquid; R is the capillary radius; ρ is the density of liquid and l is the length of capillary tube; T is the outflow time; h is the average liquid column height acting on the solution in the capillary tube, that is, $h=(h_1+$

$h_2)/2$; g is the acceleration of gravity.

The liquid in the capillary tube flows through gravity, its potential energy is used to overcome the resistance of internal friction of molecules, and at the same time, it makes the liquid itself gain kinetic energy, which leads to the low viscosity of the liquid measured in practice. If the velocity of the liquid is high, the kinetic energy consumed can be up to 20%. Therefore, Poiseuille's formula must be modified. When the liquid flows slowly, the kinetic energy consumption is small and can be ignored. At this time, h, r, l, v are constant for the same viscometer, then you have

$$\eta = k'\rho t \tag{8}$$

Considering that the measurement is usually carried out in a dilute solution of a polymer, the density, ρ, of a solution can be regarded as equal to the density, ρ_0, of a pure solvent, then the relative viscosity of the solution can be expressed as:

$$\eta_r = \frac{\eta_s}{\eta_0} = \frac{k'\rho t}{k'\rho_0 t_0} \approx \frac{t}{t_0} \tag{9}$$

Thus, the most basic measurement of polymer molecular weight by the viscosity method is t_0, t, and c. The success or failure and accuracy of the experiment depend on the accuracy of measuring the flow time of the liquid, the accuracy of preparing the solution concentration, the constant temperature of the constant temperature tank, the degree of installing the vertical position of the viscometer, the external vibration, and other factors. When determining the molecular weight of the polymer by the viscosity method, the following points should be paid attention to:

2.1 solution concentration selection.

With the increase of solution concentration, the distance between polymer molecular chains is gradually shortened. Thus the interchain forces increase. When the solution concentration exceeds a certain limitation, the relationship between $\frac{\eta_{sp}}{c}$ and $\frac{\ln\eta_r}{c}$ of the polymer solution to the concentration of c is not linear. Therefore, the usual concentration range is $\eta_r = 1.2 \sim 2.0$.

2.2 solvent selection

The solvent of a high polymer can be divided into a benign solvent and a poor solvent. In a benign solvent, polymer thread extension from the end of the chain, the chain segment density decreases, the solution $[\eta_r]$ value is bigger. On the contrary, it is difficult to dissolve in a poor solvent. When selecting solvents, the following factors need to be considered such as solubility, price, source, boiling point, toxicity, decomposition, recycling, and so on.

2.3 Capillary viscometer selection

There are two kinds of capillary viscometers in common use: Ubbelohde's viscometer and Ostwald's viscometer. Ubbelohde's viscometer is used for the determination of polymer molecular weight. For a viscometer with a volume of 5mL of 2 spheres in the Ubbelohde's viscometer as illustrated in the following figure, the flow time of the solvent is generally required to be between $t = 100 \sim 130s$.

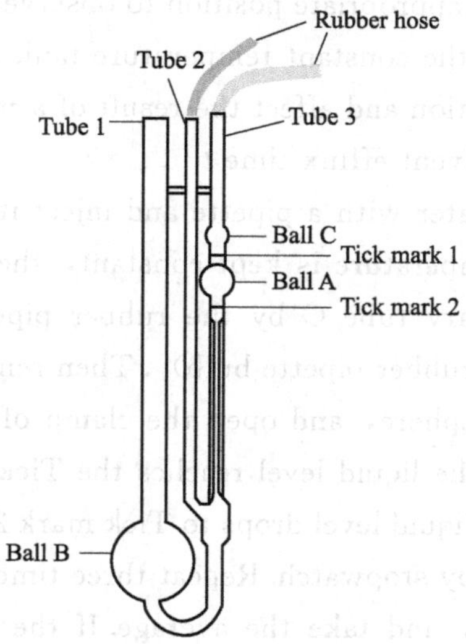

2.4 Constant temperature trough.

The temperature fluctuation directly affects the determination of solution viscosity. The national regulation is that the temperature fluctuation of the constant temperature tank is ±0.05℃ for the measurement of relative molecular mass by a viscometer.

3. Instruments and Reagents

3.1 Instruments

A set of constant temperature tanks, one Ubbelohde's viscometer, a stopwatch, one rubber pipette bulb, one volumetric flask (100mL), two pipettes (10mL), a beaker (100mL), a glass sand funnel (No. 3).

3.2 Reagents

Polyacrylic acid.

4. Procedures

4.1 Preparation of polymer solution

Weigh 0.5g polyacrylic acid into a 100mL beaker, inject about 60mL of distilled water, heat slightly to dissolve. Allow cooling to room temperature. Add 2 drops of *n*-butanol (defoamer) and transfer to 100mL volumetric flask. Add water to scale. If there are solid impurities in the solution, filter with glass sand funnel (No. 3) and set aside. Do not use filter paper to avoid mixing fibers.

4.2 Mounting viscometer

The viscometer must be dry and clean. Trace amounts of dust and grease will cause a local blockage phenomenon, and affect the flow rate of a solution in the capillary tube, which leads to large errors. So before you experiment, the viscometer must be washed thoroughly and dried in the oven. Then, a soft rubber hose is set on the upper end of the side pipe 2 and clamped to prevent air leakage. Adjust the temperature of the thermostat to 25℃. Place the viscometer vertically in a constant temperature tank, and make the ball C completely submerged in water. Place it in an appropriate position to observe the flow of liquid. The stirring speed of the stirring motor of the constant temperature tank should be adjusted properly so as not to produce violent vibration and affect the result of a measurement.

4.3 Determination of solvent efflux time t

Take 10mL of distilled water with a pipette and inject it into tube 1 of the viscometer (see Figure 2). After the temperature is kept constant, the solvent is sucked into ball A and ball C through the capillary tube C by the rubber pipette bulb (note: liquid is not allowed to be sucked into the rubber pipette bulb). Then remove the rubber pipette bulb to connect tube 2 with the atmosphere, and open the clamp of side tube 3 to let the solvent flow freely by gravity. When the liquid level reaches the Tick mark 1. Start timing by stopwatch immediately. When the liquid level drops to Tick mark 2, record the time t_0 of solvent flowing through the capillary by stopwatch. Repeat three times, each time with a difference of no more than 0.2 seconds, and take the average. If the difference is too large, check whether the capillary clogging phenomenon; Check whether the temperature of the thermostat is consistent.

4.4 Determination of efflux time of solution

After t_0 measurement, add 10mL of the prepared polyacrylic acid solution into the viscometer and pump the solution into the ball B several times with a rubber pipette bulb to make the mixture uniform. The outflow time t_1 of $c'=c/2$ was measured, and then 10mL of distilled water solution was added successively to dilute the solution with the concentration of $\frac{1}{3}c$, $\frac{1}{4}c$, and $\frac{1}{5}c$, and the outflow time t_2, t_3, and t_4 were measured, respectively (each data was repeated three times and the average was taken).

After the test, the viscometer should be washed, and then soaked in clean distilled water or inverted to dry. To remove the influence of dust, the reagent bottle and viscometer used should be kept in the bell jar, and the pipette should be covered with plastic film (do not use fiber material).

4.5 data record format and data processing

4.5.1 The experimental data were recorded in the following table

		Time				η_r	η_{sp}	$\frac{\eta_{sp}}{c}$	$\ln\eta_r$	$\frac{\ln\eta_r}{c'}$
		Measured value			Average value					
		1	2	3						
Solvent					$t_0=$					
Solution	$c'=\frac{1}{2}c$				$t_1=$					
	$c'=\frac{1}{3}c$				$t_2=$					
	$c'=\frac{1}{4}c$				$t_3=$					
	$c'=\frac{1}{5}c$				$t_4=$					

4.5.2 Drawing diagrams of $\frac{\eta_{sp}}{c'}$-c' and $\frac{\ln\eta_{sp}}{c'}$-c, and extrapolate to $c'=0$ to get $[\eta]$ value from the intercept.

4.5.3 The relative molecular weight, M_r, Of polyvinyl acid can be obtained by the formula of $[\eta]=KM^{\alpha}$

5. Experimental HSE

5.1 The key to the experiment is that the viscometer must be dry and clean until there is no water hanging on the capillary wall. The temperature of the constant temperature tank should be controlled within ±0.1℃. Polymers dissolve slowly in solvents. When preparing solutions, make sure that they are completely dissolved, and if the polymer has a flocculent, do not transfer it to the viscometer.

5.2 It must be mixed evenly to every time the solvent is added to dilution.

5.3 The viscometer must be placed vertically. Do not vibrate the viscometer during the experiment.

5.4 The viscometer is a valuable glass instrument. It must be taken care of to avoid damage during the washing, placement, and use process.

6. Questions

6.1 What is the function of branch 2 in Ubbelohde's viscometer? Can Ostwald's viscometer be used in this experiment?

6.2 How does the capillary diameter of the viscometer affect the experimental results?

6.3 When measuring viscosity, the viscometer must be vertical and put into a constant temperature tank. Why?

6.4 Is it ok to measure relative viscosity without using the same viscometer?

6.5 Evaluate the advantages and disadvantages of the viscosimetry for the determination of polymer molecular weight and point out the factors that affect the accurate determination results

7. Discussion

7.1 There are two kinds of capillary viscometers in common use: one is the three-tube viscometer, which is Ubbelohde's viscometer used in this experiment. The characteristic is that the outflow time of the solution is independent of the volume of the liquid to be tested, so the concentration of the liquid to be tested can be changed by adding solvent or solution to the viscometer. The other is a two-tube viscometer, which is the Ostwald's viscometer. Since the outflow time of the liquid is related to the height of the solution added to the viscometer, the volume of the standard liquid and the liquid to be measured must be the same. Considering that $\frac{mV}{8\pi lt}$ can be ignored, appropriate capillary length, diameter, and e-sphere size should be selected. Appropriate capillary length, diameter, and e-sphere size should be selected so that the outflow time is more than 100s, preferably around 120s. But the capillary diameter should not be too fine, which is easy to block the viscosity meter. After use, the viscometer should be cleaned immediately to prevent polymer bonding and even blockage of the capillary aperture. After cleaning, the viscometer should be filled with distilled water, which will keep out dust.

7.2 The unit of characteristic viscosity and the unit of concentration are reciprocal to each other. However, the units used in the literature and the experimental textbooks are not the same, resulting in the order of magnitude difference in the parameter K in the formula $[\eta] = K\overline{M}^{\alpha}$. Parameter K has the same units as $[\eta]$, whereas parameter α is dimensionless. For the same polymer, the values of K and α are different at different temperatures and different solvent conditions. Attention should be paid when using.

7.3 The behavior of polymer chains in solution will affect the determination of $[\eta]$.

7.3.1 Polyelectrolyte behavior, that is, the side groups of some polymer chains can be ionized, and the ionized polymer chains have mutual repulsion. With the decrease of c, η_{sp}/c increases abnormally. A small number of small molecules of an electrolyte can be added as inhibitors to inhibit the ion effect.

7.3.2 Some polymers will degrade in solution, making the results of $[\eta]$ and M lower, which can be inhibited by adding a small amount of antioxidant.

7.4 Influence factors of inverse linearity of η_{sp}/c-c and η_r/c-c.

7.4.1 temperature fluctuation

In general, for different solvents and polymers, temperature fluctuations have different effects on viscosity. The relationship between solution viscosity and temperature can be expressed by Andraole's equation $\eta = Ae^{B/(RT)}$, where A and B are constant for A given polymer and solvent. R is the gas constant. Therefore, it requires a good temperature control precision.

7.4.2 concentration of the solution

With the increase of the concentration, the distance between polymer molecular chains is gradually shortened, so the intermolecular chain interaction force increases. When the concentration exceeds a certain limit, the relationship between η_{sp}/c or ln η_r/c of the polymer solution and c is not linear. The concentration range of $\eta_r = 1.2 \sim 2.0$ is usually used.

7.4.3 the flow time is affected by vertical changes of the capillary tube and local blockage of particulate impurities during the determination.

7.5 Approximate treatment of abnormal phenomena in viscosity measurement.

In the process of specific viscosity measurement, the abnormal phenomenon of drawing and drawing extrapolation do not intersect at a single point on the vertical axis sometimes occurs without careless operation. At this time, the characteristic viscosity, $[\eta]$, of the polymer solution should be obtained based on the relationship between η_{sp}/c-c.

实验十三　酶催化分光光度法测定过氧化氢

一、实验目的

1. 了解 TMB-HRP-H_2O_2 显色体系的基本原理。

2. 掌握紫外-可见分光光度计的操作技术和有关参数的测定方法。

二、实验原理

目前检测过氧化氢的方法主要有以下几类：分光光度法、化学发光法、荧光光度法、色谱法、电化学分析法、氧化还原滴定法等，这些方法各有优缺点。氧化还原滴定法（如 $KMnO_4$ 法等）测定过氧化氢需要目视来测定终点，该方法操作简单，适用于常量分析，但是方法灵敏度低。分光光度法因分析操作简单，不需要复杂的设备，选择性好，在各个领域得到广泛应用。而酶催化分光光度法测定过氧化氢是以 TMB 为显色剂，HRP 催化 H_2O_2 氧化 TMB 酶促反应，在 HRP 的催化作用下，TMB 被 H_2O_2 氧化，溶液从无色转变为蓝色，该酶促反应经 H_2SO_4 终止后，最终产物使溶液颜色显示为黄色，可用紫外-可见分光光度计记录该产物在 450nm 处的吸收值。上述反应可用下述反应方程式表示：

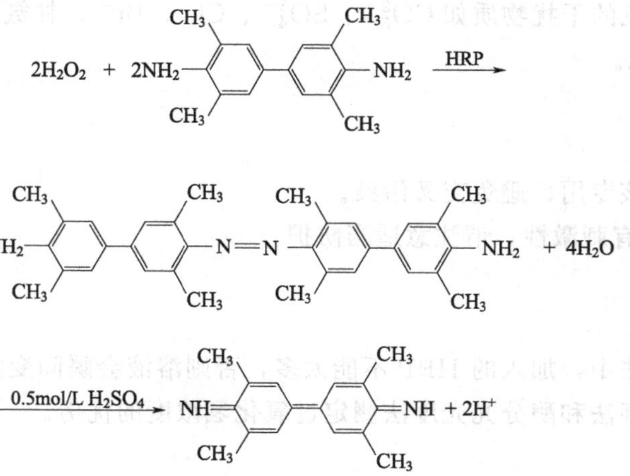

三、仪器与试剂

1. 仪器

奥立龙 CHN060 型酸度计，pH 玻璃电极，紫外-可见分光光度计，恒温水槽，分析天平，移液管（25mL），容量瓶（100mL×2），比色管（10mL），移液枪（1000μL），1.5mL 塑料带盖离心管。

2. 试剂

过氧化氢标准溶液（0.01mol/L），四甲基联苯胺（TMB），无水乙醇，辣根过氧化物酶（HRP，250u/mg）（4℃下保存），25% HCl，$Na_2HPO_4 \cdot 12H_2O$，$KH_2PO_4 \cdot 2H_2O$，NaCl，0.5mol/L H_2SO_4，Na_2CO_3，Na_2SO_4，NaCl，NaBr，甘氨酸，医用双氧水样品，0.5mol/L KCl。

四、实验步骤

1. 稀释得到过氧化氢浓度分别为 0.1mmol/L、0.4mmol/L、1mmol/L、3mmol/L、5mmol/L 的标准溶液，现配现用。

2. 准确称取 TMB 0.0962g，用 50mL 无水乙醇溶解，用水定容至 100mL，其浓度为 4mmol/L。

3. 称取 0.8g NaCl、0.144g Na_2HPO_4 和 0.024g KH_2PO_4，溶于 80mL 蒸馏水中，用

HCl调节溶液的pH值至7.4，最后加蒸馏水定容至100mL，即可得到pH＝7.4的0.01mol/L磷酸盐缓冲溶液。

4. 在10mL比色管中依次加入2.0mL pH＝7.4的磷酸盐缓冲溶液，0.4mL 4mmol/L TMB溶液，1.0mL不同浓度的过氧化氢标准溶液，1.0mL 5μmol/L HRP溶液，用水定容，混合均匀后，37℃下避光温育10min。

5. 反应用1mL 0.5mol/L H_2SO_4终止，记录450nm处吸光度A。以水为参比，同时测定空白溶液的吸光度A_0，计算吸光度差值$\Delta A = A - A_0$。以ΔA对过氧化氢的浓度制作标准工作曲线，得标准曲线方程及相关系数。

6. 取不同型号的医用双氧水（主要成分为过氧化氢）2mL，用0.5mol/L氯化钾溶液稀释至500mL，按实验方法进行测定，记录吸光度差值$\Delta A = A - A_0$，查标准曲线，计算样品中过氧化氢的浓度，同时进行加标回收实验。

7. 考察一些常见的干扰物质如CO_3^{2-}、SO_4^{2-}、Cl^-、Br^-、甘氨酸对TMB-HRP-H_2O_2显色系统的干扰情况。

五、实验HSE

1. 实验用品应该专用，避免交叉污染。
2. TMB对人体有刺激性，请注意适当防护。

六、思考题

1. 酶分光光度法中，加入的HRP不能太多，否则溶液会瞬间变黄，请解释原因。
2. 比较高锰酸钾法和酶分光光度法测定过氧化氢浓度的优劣。

Experiment 13 Determination of hydrogen peroxide by enzyme catalytic spectrophotometry

1. Objectives

1.1 Understand the basic principles of the TMB-HRP-H_2O_2 color rendering system.

1.2 Master the operating techniques of the UV-vis spectrometer and the methods for determining the relevant parameters.

2. Principle

At present, there are mainly the following methods to detect hydrogen peroxide: including spectrophotometry, chemiluminescence, fluorescence spectrometry, chromatography, electrochemical analysis, redox titration, and so on. Each of these methods has advantages and disadvantages. Hydrogen peroxide is determined by redox titration (for example, $KMnO_4$ method), and the endpoint is determined by the visual method. The method is simple and suitable for constant analysis, but the sensitivity is low. Spectrophotometry is widely used in various fields because of its simple operation, no complicated equipment, and good selectivity. For example, the method of hydrogen peroxide determination by enzyme catalytic spectrophotometry. In the process, TMB is used as a chromogenic agent and HRP catalyzes the enzymatic reaction of TMB oxidized by H_2O_2. Under the catalysis of HRP, TMB is oxidized by H_2O_2, and the solution changes from colorless to blue. After the termination of the enzymatic reaction by H_2SO_4, the final product turns the solution yellow. The absorption value of the product at 450nm can be recorded by the UV-vis spectrometer. The above reaction can be expressed by the following reaction equation:

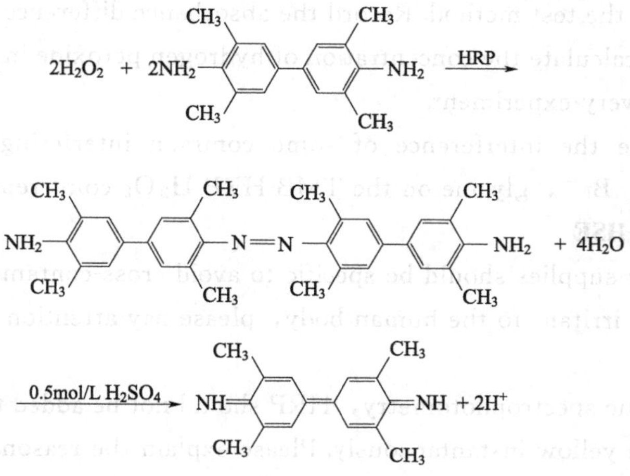

3. Instruments and reagents

3.1 Instruments

Orion CHN060 acidimeter, pH glass electrode, UV-vis spectrophotometer, thermostatic water bath, analytical balance, pipette (25mL), capacity bottle (100mL×2), colorimetric tube (10mL), 1.5mL plastic capped centrifugal tube.

3.2 Reagents

Hydrogen peroxide standard solution (0.01mol/L), tetramethylbenzidine (TMB), absolute ethanol, horseradish peroxidase (HRP, 250u/mg) (stored at 4℃), 25% HCl, $Na_2HPO_4 \cdot 12H_2O$, $KH_2PO_4 \cdot 2H_2O$, NaCl, 0.5mol/L H_2SO_4, Na_2CO_3, Na_2SO_4, NaCl, NaBr, glycine, medical hydrogen peroxide sample, 0.5mol/L KCl.

4. Procedures

4.1 The hydrogen peroxide standard solution with the concentration of 0.1mmol/L, 0.4mmol/L, 1 mmol/L, 3 mmol/L, 5 mmol/L were obtained by dilution.

4.2 Accurately weigh 0.0962g TMB, dissolve it in 50mL absolute ethanol and dilute to 100mL with water at a concentration of 4 mmol/L.

4.3 Take 0.8g NaCl, 0.144g Na_2HPO_4, and 0.024g KH_2PO_4 and dissolve them in 80mL distilled water. Adjust the pH value of the solution to 7.4 with HCl, and finally dilute to 100mL with distilled water to obtain 0.01mol/L phosphate buffer solution with pH=7.4.

4.4 Add 2.0mL of pH 7.4 phosphate buffer solution, 0.4mL of 4mmol/L TMB solution, 1.0mL of hydrogen peroxide standard solution with different concentrations and 1.0mL of 5μmol/L HRP solution into a 10mL colorimetric tube. Then constant volume with water and mix well, and incubate at 37℃ for 10 min in the dark.

4.5 The reaction is terminated with 1mL 0.5mol/L H_2SO_4 and the absorbance A at

450nm is recorded. The absorbance of blank solution A_0 is simultaneously determined with water as a reference, and the absorbance difference value $\Delta A = A - A_0$ is calculated. The standard working curve of ΔA with hydrogen peroxide concentration is made, and the standard curve equation and correlation coefficient are obtained.

4.6 Take 2mL of different types of medical hydrogen peroxide (the main component is hydrogen peroxide), dilute to 500mL with 0.5mol/L potassium chloride solution, and measure according to the test method. Record the absorbance difference $\Delta A = A - A_0$, check the standard curve, calculate the concentration of hydrogen peroxide in the sample, and carry out the spike recovery experiment.

4.7 Investigate the interference of some common interfering substances such as CO_3^{2-}, SO_4^{2-}, Cl^-, Br^-, glycine on the TMB-HRP-H_2O_2 color rendering system.

5. Experimental HSE

5.1 Laboratory supplies should be specific to avoid cross-contamination.

5.2 TMB is an irritant to the human body, please pay attention to proper protection.

6. Questions

6.1 In enzymatic spectrophotometry, HRP should not be added too much. Otherwise, the solution will turn yellow instantaneously. Please explain the reason.

6.2 To compare the advantages and disadvantages of the potassium permanganate method and enzymatic spectrophotometry in the determination of hydrogen peroxide concentration.

实验十四　磷酸的电位滴定

一、实验目的

1. 掌握电位滴定的原理和操作技术。
2. 学会用电位滴定法测定弱酸的电离常数。
3. 学会 pH 计及玻璃电极的使用方法。

二、实验原理

用一般的容量分析方法来滴定磷酸，由于磷酸是多元弱酸，存在分步电离：

$$H_3PO_4 \rightleftharpoons H_2PO_4^- + H^+ \qquad pK_{a_1} = 2.12$$

$$H_2PO_4^- \rightleftharpoons HPO_4^{2-} + H^+ \qquad pK_{a_2} = 7.20$$

$$HPO_4^{2-} \rightleftharpoons PO_4^{3-} + H^+ \qquad pK_{a_3} = 12.36$$

化学计量点时，pH 突跃比较小，而且要用两种指示剂，观察指示剂的颜色变化比较困难，容易造成误差。用电位滴定法滴定磷酸，不需用指示剂，用$\frac{\Delta pH}{\Delta V}$-$V$ 图，可准确地确定滴定终点。根据 pH-V 曲线可以求出磷酸的 pK_{a_1}、pK_{a_2}。这是一般容量分析做不到的。

电位滴定确定终点的方法如图 1 所示。

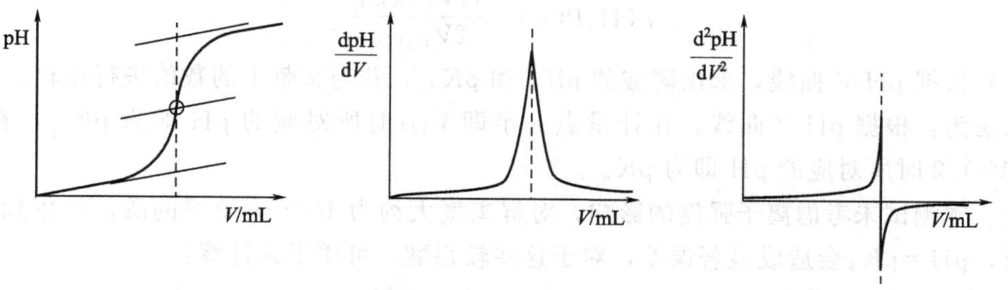

(a) 三切线法图　(b) 一级微商法图　(c) 二级微商法图

图 1　电位滴定确定终点的方法

三、仪器和试剂

1. 仪器

奥立龙 CHN060 型酸度计，pH 玻璃电极，饱和甘汞电极，电磁搅拌器，50mL 碱式滴定管 1 支，10mL 移液管 1 支，150mL 烧杯 2 只。

2. 试剂

0.1mol/L NaOH 溶液，0.1mol/L H_3PO_4 溶液，pH 标准溶液，0.1mol/L HCl 标准溶液，酚酞指示剂。

四、实验步骤

1. 将仪器接通电源，打开电源开关，预热半小时。

2. 按照仪器使用说明安装电极，调节零点，用混合磷酸盐标准缓冲溶液校正仪器，校正后，洗净电极。

3. 称取 0.87g NaOH 固体，溶于 200mL 蒸馏水，配成 NaOH 溶液（0.1mol/L NaOH 溶液）。

4. 准确移取 25mL 0.1mol/L HCl 标准溶液于 250mL 锥形瓶，滴加 1～2 滴酚酞指示剂，将 NaOH 溶液装入滴定管，进行酸碱滴定，平行滴定 2 次，求 NaOH 的准确浓度。

5. 准确移取 0.1mol/L 磷酸溶液 10mL 于 150mL 烧杯中，用蒸馏水稀释至 80mL，放入电极及搅拌磁子，开动电磁搅拌器，测量其 pH 值。安装仪器、滴定过程中搅拌溶液时，要防止碰破玻璃电极。

6. 慢慢滴加 0.1mol/L 的 NaOH 溶液（应新配制，不含 CO_3^{2-}），每滴加 2mL 左右测量一次 pH 值，准确记录滴定管读数。当溶液 pH 值大于 3 以后，每滴加 0.2mL 左右测量一次 pH 值。当溶液 pH 值大于 6 以后，每滴加 1mL 左右测量一次 pH 值。当 pH 值大于 7.5 以后，每滴加 0.2mL 左右测量一次 pH 值。当 pH 值大于 11 以后，滴定即可停止。滴定剂加入后，要充分搅拌溶液，停止时再测定 pH 值，以得到稳定的读数。

7. 重复上述滴定一次。测完后要将电极置于水中复原。

8. 附实验数据处理要求

（1）根据滴定所用的 NaOH 溶液的体积 V 和相应的 pH 值，绘出 pH-V 及 $\frac{\Delta pH}{\Delta V}$-$V$ 曲线。

由第二化学计量点的 NaOH 体积及 NaOH 的浓度算出磷酸的准确浓度。

$$c(H_3PO_4)=\frac{(cV_2)_{NaOH}}{2V_{H_3PO_4}}$$

（2）根据 pH-V 曲线，求出磷酸的 pK_{a_1} 和 pK_{a_2}，并与文献上的数值进行比较。

求法为：根据 pH-V 曲线，在计量点一半即 $V_{1/2}$ 时所对应的 pH 即为 pK_{a_1}，在 $V_1+(V_2-V_1)/2$ 时所对应的 pH 即为 pK_{a_2}。

注：本测试未考虑离子强度的影响，对解离度大约为 10^{-3}～10^{-2} 的酸，使用到计量点一半时，pH＝pK_a 会造成显著误差，对于这些较强酸，可用下式计算：

$$pH=pK_a-\lg\frac{c_{酸}-[H^+]}{c_{酸}+[H^+]}$$

式中，$c_{酸}$ 是在计量点一半时共轭酸的浓度，mol/L。

五、实验 HSE

NaOH 溶液有一定的腐蚀性，在使用时必须防止被它腐蚀。如果不慎将碱液沾到皮肤上，要立刻用大量的水冲洗，再涂上稀硼酸溶液对碱液进行中和，从而减轻碱液对皮肤的腐蚀。

六、思考题

1. 用 NaOH 滴定 H_3PO_4，第一化学计量点和第二化学计量点所消耗的 NaOH 体积理应相等，为什么实际上并不相等？

2. 如何根据滴定弱碱的数据求它的 K_b？

3. 磷酸的第三级电离常数 K_{a_3} 可以从滴定曲线上求得吗？

Experiment 14 Potentiometric titration of phosphoric acid

1. Objectives

1.1 Master the principles and operation technique of potentiometric titration.

1.2 Learn the determination of ionization constant of a weak acid by potentiometric titration.

1.3 Learn the usage of pH meter and glass electrode.

2. Principle

When the phosphoric acid is titrated by the general volumetric analysis method, the pH jump at the stoichiometric point is relatively small, and it is difficult to observe the color change of the indicator with two kinds of indicators, which is easy to cause errors. As phosphoric acid is a weak acid, there is step-by-step ionization.

$$H_3PO_4 \rightleftharpoons H_2PO_4^- + H^+ \qquad pK_{a_1}=2.12$$

$$H_2PO_4^- \rightleftharpoons HPO_4^{2-} + H^+ \qquad pK_{a_2}=7.20$$

$$HPO_4^{2-} \rightleftharpoons PO_4^{3-} + H^+ \quad pK_{a_3} = 12.36$$

Phosphoric acid is titrated by potentiometric titration without an indicator, the titration endpoint can be accurately determined through the relation between $\frac{\Delta pH}{\Delta V}$ and V. According to the curve of pH-V, pK_{a_1}, and pK_{a_2} of phosphoric acid can be calculated. This cannot be done by general volumetric analysis.

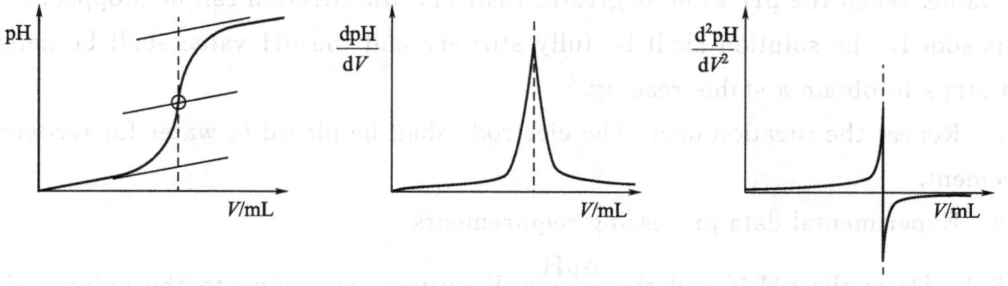

(a) three tangent method (b) the first derivative method (c) the second derivative method

Figure. 1 The methods for determining the endpoint by potentiometric titration

3. Instruments and reagents

3.1 Instruments

Alalis pH 400, glass electrode, electromagnetic stirrer, base buret (50mL × 1), pipette (10mL×1), beaker (150mL×2), conical flask (250mL×2).

3.2 Reagents

NaOH (0.1mol/L), H_3PO_4 (0.1mol/L), pH standard solution, phenolphthalein indicator, HCl standard solution (0.1mol/L).

4. Procedures

4.1 Turn the instrument on, turn on the power switch, and preheat it for half an hour.

4.2 Install the electrodes according to the instructions, adjust the zero point, calibrate the instrument with the phosphate standard buffer solution, and then clean the electrode after calibration.

4.3 NaOH solid (0.87g) was weighed and dissolved in 200mL of distilled water to prepare a NaOH solution (0.1mol/L NaOH).

4.4 HCl standard solution (0.1mol/L, 25mL) was accurately added in a conical flask (250mL), 1～2 drops of phenolphthalein indicator were added, the NaOH solution was loaded into the base burette, then acid-base titration was performed. The titration processes were conducted parallel titration twice to obtain the accurate concentration of NaOH.

4.5 H_3PO_4 (0.1mol/L, 10mL) was accurately assimilated in a beaker (150mL), and the volume of H_3PO_4 was diluted to 80mL. Electrode and Stirring magneton were put into the solution of H_3PO_4. Open electromagnetic stirrer, and measure its pH value. When installing the instrument and stirring the solution during titration, the glass electrode shall not be broken.

4.6 Fresh NaOH solution was added in the above solution drop by drop. When about 2mL of NaOH was added, the pH value was measured once, and the burette reading was accurately recorded. When the pH value is bigger than 3, every drop (~0.2mL) was added, and the pH value was measured once. When the pH value is greater than 6, the volume of NaOH (~1mL) was added, and the pH value was measured once. When the pH value is greater than 7.5, add the volume of NaOH (~0.2mL) for each drop to measure the pH value. When the pH value is greater than 11, the titration can be stopped. After the titrant is added, the solution shall be fully stirred, and the pH value shall be determined when it stops to obtain a stable reading.

4.7 Repeat the titration once. The electrode shall be placed in water for recovery after measurement.

4.8 Experimental data processing requirements

4.8.1 Draw the pH-V and the $\frac{\Delta pH}{\Delta V}-V$ curves, according to the volume V of the NaOH solution used for titration and the corresponding pH value. The accurate concentration of phosphoric acid is calculated from the volume of NaOH at the second stoichiometric point and the concentration of NaOH.

$$c(H_3PO_4)=\frac{(cV_2)_{NaOH}}{2V_{H_3PO_4}}$$

4.8.2 According to the curve of pH-V, pK_{a_1} and pK_a were obtained. Briefly, pK_{a_1} is at the pH corresponding to half of the metering point ($V_{1/2}$), pK_{a_2} is at the pH corresponding to the volume at $V_1+(V_2-V_1)/2$. These values were compared with references.

PS: In this test, the effect of ionic strength was not considered. When dissociation constants of acids were about 10^{-3}~10^{-2}, when pH=pK_a was used at the halfway point, significant errors would be caused. For these strong acids, the following equation can be used to calculate:

$$pH=pK_a-\lg\frac{c_a-[H^+]}{c_a+[H^+]}$$

Note: c_a was the concentration of acid or base at the halfway point, unit: mol/L.

5. Experimental HSE

NaOH solution has certain corrosiveness, so it must be prevented from corroding when it is used. If it is accidentally touched on the skin, wash it immediately with a large amount of water, and then apply the dilute boric acid solution to neutralize the lye, to reduce the corrosion of the lye on the skin.

6. Questions

6.1 When H_3PO_4 is titrated with NaOH, the volume of NaOH consumed at the first and the second stoichiometric points should be equal. Why is it not equal in fact?

6.2 How to get K_b of weak base according to data of titrating?

6.3 Can the third ionization constant K_{a_3} of H_3PO_4 be obtained from the titration curve?

实验十五　对乙酰氨基酚的电化学反应机理及其浓度的测定

一、实验目的

1. 了解循环伏安法的基本原理及应用。
2. 掌握循环伏安法的实验技术和有关参数的测定方法。
3. 学习用循环伏安法研究乙酰氨基酚的电化学氧化机理的方法。

二、实验原理

循环伏安法（Cyclic voltammetry，CV）是在固定面积的工作电极和参比电极之间加上对称的三角波扫描电压，记录工作电极上得到的电流与施加电位的关系曲线，即循环伏安图。从伏安图的波形、氧化还原峰电流的数值及其比值、峰电位等可以判断电极反应机理。可用循环伏安法来检测物质的氧化还原电位，考察电化学反应的可逆性和反应机理，判断产物的稳定性，研究活性物质的吸附和脱附现象，也可用于反应速率的半定量分析等。循环伏安法在工作电极上施加一个线性变化的循环电压，记录工作电极上得到的电流与施加电压的关系曲线，对溶液中的电活性物质进行分析。由于施加的电压为三角波（图 1），这种方法也称为三角波线性扫描极谱法。

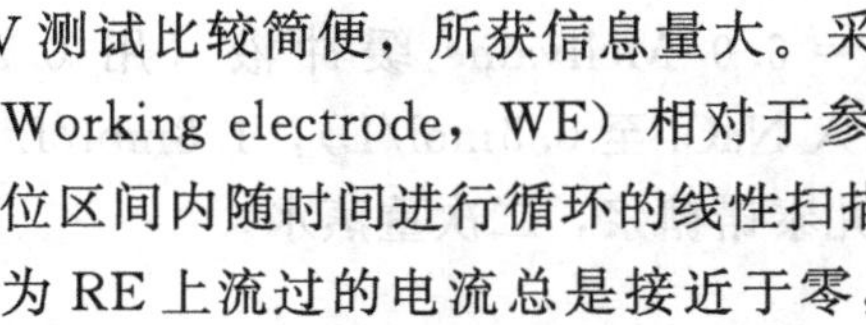

图 1　施加电压与时间的关系

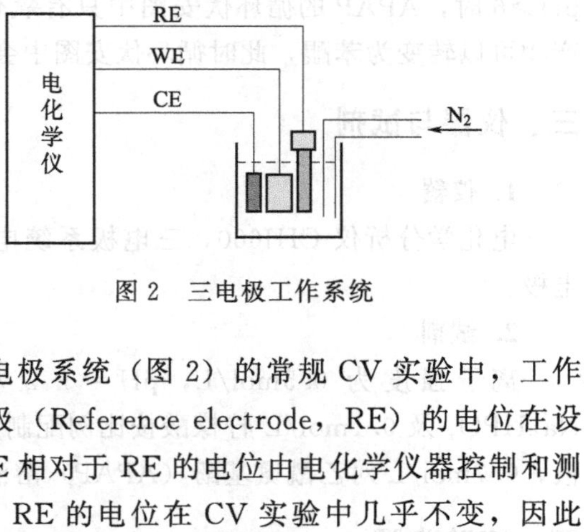

图 2　三电极工作系统

CV 测试比较简便，所获信息量大。采用三电极系统（图 2）的常规 CV 实验中，工作电极（Working electrode，WE）相对于参比电极（Reference electrode，RE）的电位在设定的电位区间内随时间进行循环的线性扫描，WE 相对于 RE 的电位由电化学仪器控制和测量。因为 RE 上流过的电流总是接近于零，所以 RE 的电位在 CV 实验中几乎不变，因此 RE 是实验 WE 电位测控过程中的稳定参比。若忽略流过 RE 上的微弱电流，则实验体系的电解电流全部流过由 WE 和对电极（Counter electrode，CE）组成的串联回路。WE 和 CE 间的电位差可能很大，以保证能成功地施加如上所述设定的 WE 电位（相对于 RE）。CE 也常称为辅助电极（Auxiliary electrode，AE）。

典型的循环伏安图如图 3 所示，该图是在 KCl 电解质溶液中，$K_3Fe(CN)_6$ 在玻碳电极上反应得到的结果。循环伏安图的几个重要参数为：阳极峰电流（i_{pa}）、阴极峰电流（i_{pc}）、阳极峰电位（E_{pa}）、阴极峰电位（E_{pc}）。对于可逆氧化还原电对，阴阳极峰电位的差值即

$$\Delta E = E_{pa} - E_{pc} \approx 0.059/n \quad (1)$$

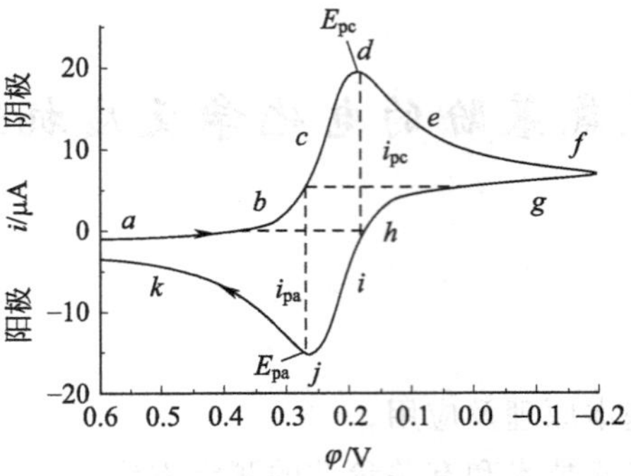

图 3　KCl 电解质溶液中 $K_3Fe(CN)_6$ 的循环伏安图

对可逆体系的正向峰电流，由 Randles-Savcik 方程可表示为：

$$i_p = 2.69 \times 105 n^{3/2} A D^{1/2} V^{1/2} c \tag{2}$$

式中，i_p 为峰电流，A；n 为电子转移数；A 为电极面积，cm^2；D 为扩散系数，cm^2/s；V 为扫描速度，V/s；c 为浓度，mol/L。

由此可见，i_p 与 $V^{1/2}$ 和 c 呈线性关系。

在可逆电极反应过程，

$$i_{pa}/i_{pc} \approx 1 \tag{3}$$

对一个简单的电极反应过程，式(1) 和式(3) 是判别电极反应是否为可逆体系的重要依据。

对乙酰氨基酚 APAP 电化学氧化机理：在 pH≤6 的缓冲溶液中，APAP 可在电极表面被迅速氧化，由于 H^+ 参与电化学氧化，APAP 氧化峰电位应随溶液 pH 的改变而变化；在 pH>6 时，APAP 的循环伏安图中只有氧化峰，没有还原峰。在酸性极大的溶液中，中间产物可以转变为苯醌，此时循环伏安图中会出现苯醌的还原峰。

三、仪器与试剂

1. 仪器

电化学分析仪 CHI660，三电极系统电解池，玻碳电极，铂辅助电极，Ag/AgCl 参比电极。

2. 试剂

离子强度为 0.5mol/L，pH = 2.2 和 pH = 6.0 Mcllvaine 缓冲液（用 0.2mol/L Na_2HPO_4 及 0.1mol/L 柠檬酸按比例配制并各加入 NaCl 至 0.5mol/L），1.8mol/L 硫酸溶液，0.7mol/L 对乙酰氨基酚（APAP）溶液，小儿泰诺糖浆，二次重蒸水。

四、实验步骤

1. 各配制 10mL 底液分别为 pH=2.2、pH=6.0 的 Mcllvaine 缓冲液及 1.8mol/L 硫酸的 3mmol/L APAP 溶液，并配制底液为 pH = 2.2 的 Mcllvaine 缓冲液、浓度分别为 0.1mmol/L、0.4mmol/L、1mmol/L、5mmol/L 的 APAP 标准溶液，包括上述配制的 pH=2.2 的 Mcllvaine 的 3mmol/L APAP 标准溶液，共为 5 个标准溶液。

2. 装上三电极系统，将电极引线分别接入循环伏安仪，并接上记录仪等。

3. 按浓度从低到高的顺序分别将三电极系统插入 5 个 pH=2.2 的 APAP 标准溶液中，以 40mV/s 的速率作循环伏安扫描。分别记录氧化峰的峰电位及峰高值。将峰高值对浓度作线性回归分析，得标准曲线方程及相关系数。做循环伏安法实验要注意，每次做完实验都要仔细打磨电极。防止电极表面在实验过程中吸附其他物质，覆盖电极表面，减小电极接触面积，影响实验数据，产生误差。同时各电极的接口要避免互相接触以免发生短路。

4. 按上述相同实验方法测定小儿泰诺糖浆的5倍稀释液（用pH＝2.2的Mcllvaine缓冲液稀释），记录氧化峰的峰高值。查标准曲线，计算样品中的APAP的浓度，并与药品包装盒中APAP的标志值对照。

5. 将底液各为pH＝2.2、pH＝6.0 Mcllvaine缓冲液及1.8mol/L硫酸溶液的3mmol/L APAP溶液分别以40mV/s和250mV/s的速率作循环伏安扫描，记录循环伏安图，观察在不同pH下，不同扫描速率下的氧化峰及还原峰情况。证实实验原理中提出的APAP电化学氧化机理。

6. 实验数据处理要求

（1）根据pH＝2.2条件下，不同浓度（0.1mmol/L、0.4mmol/L、1mmol/L、3mmol/L、5mmol/L）乙酰氨基酚溶液的循环伏安图，绘出浓度-氧化峰电流的曲线关系。

（2）根据样品的循环伏安图氧化峰的大小，查标准曲线，计算样品中APAP的浓度，并与药品包装盒中APAP的标志值对照。

（3）根据不同pH（pH＝2.2、pH＝6.0、1.8mol/L硫酸）溶液中乙酰氨基酚的循环伏安图，讨论氧化峰、还原峰的电流电位值与溶液pH的关系。证实实验原理中提出的APAP电化学氧化机理。

五、实验HSE

硫酸对皮肤、黏膜等组织有强烈的刺激和腐蚀作用，使用时注意防护。

六、思考题

1. 对比pH＝2.2和强酸介质的扫描图谱，可以看出在强酸循环伏安图中新出现一个还原峰，请给出解释。

2. 在pH＝6.0的体系中，根据不同扫描速率的ΔE_p判断APAP在电极表面的电化学反应是否为可逆过程。

3. 相同扫描速率下，pH＝6.0、pH＝2.2的Mcllvaine缓冲液和强酸介质中，为什么APAP的氧化电位值会随着酸度的改变而发生移动？

Experiment 15 Study on the electrochemical reaction mechanism of acetaminophen and determination of its concentration

1. Objectives

1.1 Understand the basic principles and applications of Cyclic voltammetry.

1.2 Master the experimental techniques and methods for determining the related parameters of cyclic voltammetry.

1.3 Study on the method of electrochemical oxidation mechanism of acetaminophen by Cyclic voltammetry.

2. Principle

The Cyclic voltammetry (CV) which adds a symmetrical triangular wave scan voltage between a fixed-area working electrode (WE) and a reference electrode (RE). It can record the relationship curve between the current obtained on the working electrode and the applied potential. The relationship curve is called a cyclic voltammogram. The electrode reaction mechanism can be judged from the waveform of the voltammogram. The value of redox peak current, its ratio, and peak potential can have the same effect. It can be used to detect the redox potential of a substance, investigate the reversibility of the electrochemical reaction and the reaction mechanism, determine the stability of the product, study the adsorption and desorption of the active substance. Besides, it can also be used for semi-quantitative analysis of the reaction rate. The CV which adds a linearly varying cyclic voltage across the working electrode. It can record the current vs. applied voltage on the working electrode and analyze the electroactive species in the solution. Since the applied voltage is a triangular wave (Fig. 1), this method is also called triangular wave linear scanning polarography.

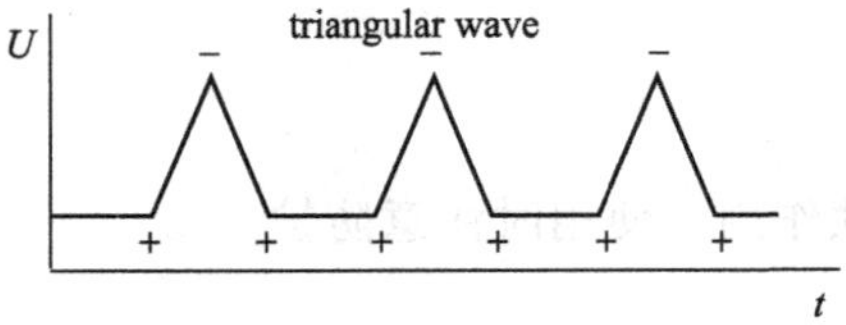

Fig. 1 Potential-time excitation signal

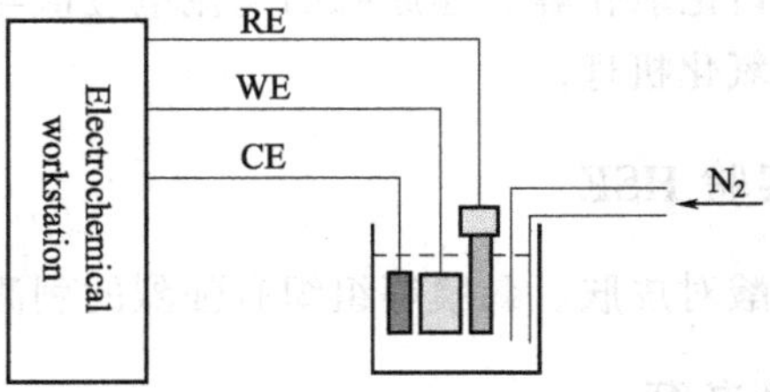

Fig. 2 Three-electrode working system

Although the CV test is relatively simple, a large amount of information is obtained. A three-electrode system is using in a conventional CV experiment (Fig. 2), the potential of the WE relative to RE is cyclically linear scanned over time within a set potential interval. The potential of WE relative to RE is controlled and measured by electrochemical instruments. Since the current flowing through the RE is always close to zero, the potential of the RE is almost constant in the CV experiment. Therefore, the RE is a stable reference in the process of measuring and controlling the WE potential. If the weak current flowing through the RE is neglected, the electrolysis current of the experimental system flows through the series circuit which is made up of WE and the counter electrode (CE). The different potential between WE and CE may be large to ensure the WE potential (relative to RE) setting successfully. CE is generally called the Auxiliary Electrode (AE).

A typical CV is shown in Figure 3. This figure is the result of the reaction of $K_3Fe(CN)_6$ on a glassy carbon electrode in a KCl electrolyte solution. The following are some important parameters of the CV: Anode peak current (i_{pa}), Cathode peak current (i_{pc}), Anode peak potential (E_{pa}), Cathode peak potential (E_{pc}).

For the reversible oxidation-reduction couples, the difference between the anodic and

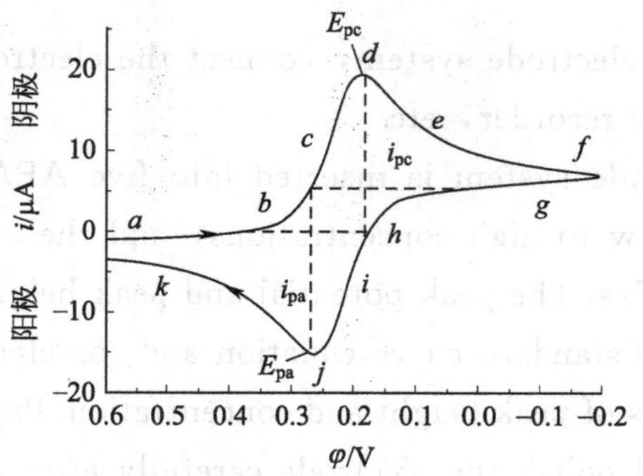

Fig. 3 Cyclic voltammogram of $K_3Fe(CN)_6$ in KCl electrolyte solution

cathodic peak potentials as follows:

$$\Delta E = E_{pa} - E_{pc} \approx 0.059/n \tag{1}$$

The positive peak current for a reversible system is given by Randles-Savcik equation:

$$i_p = (2.69 \times 10^5) n^{3/2} A D^{1/2} V^{1/2} c \tag{2}$$

where i_p is the peak current (in A), n is the transfer number of electrons, A is the electrode area (in cm^2), D is the diffusion coefficient (in cm^2/s), V is the scan rate (in V/s), and c is the concentration (in mol/cm^3). Thus, i_p has a linear relationship with $V^{1/2}$ and c. In a reversible electrode reaction process,

$$i_{pa}/i_{pc} \approx 1 \tag{3}$$

For a simple electrode reaction process, equations (1) and (3) are an important reference for determining whether the electrode reaction is reversible.

The electrochemical oxidation mechanism of APAP: APAP can be rapidly oxidized on the electrode surface in a buffer solution with pH$\leqslant$6. Since H^+ is involved in electrochemical oxidation, the oxidation peak potential of APAP should be changed with the pH of the solution. ; The CV of APAP has no reduction peak in a solution with pH $>$ 6. The intermediate product can be converted to benzoquinone in strongly acidic solutions, and the reduction peak of benzoquinone appears in a CV at this time.

3. Instruments and reagents

3.1 Instruments

Electrochemical analyzer CHI660, three-electrode system electrolytic cell, glassy carbon electrode, platinum auxiliary electrode, Ag/AgCl reference electrode.

3.2 Reagents

Ionic strength 0.5, pH = 2.2 and pH = 6.0 Mcllvaine buffer solution (prepared with 0.2mol/L Na_2HPO_4 and 0.1mol/L citric acid in proportion followed by adding 0.5mol/L NaCl), 1.8mol/L Sulfuric acid solution, 0.7mol/L acetaminophen solution, Children's Tylenol syrup, redistilled water.

4. Procedures

4.1 Prepare10mL of 3mmol/L APAP solution in 1.8mol/L sulfuric acid or pH = 2.2 or pH = 6.0 Mcllvaine buffer . Prepare the standard solutions of APAP in pH = 2.2 acetaminophens with different concentrations of 0.1mmol/L, 0.4mmol/L, 1mmol/L and 5mmol/L, respectively.

4.2 Install the three-electrode system, connect the electrode leads to the cyclic voltammeter and connect to the recorder, etc.

4.3 The three-electrode system is inserted into five APAP standard solutions with pH=2.2 in the order of low to high concentrations, and the cyclic voltammetry was performed at the rate of 40mV/s. The peak potential and peak height of the oxidation peak are recorded, respectively. The standard curve equation and correlation coefficient are obtained by linear regression analysis of peak height and concentration. Pay attention to the cyclic voltammetry experiment, and polish the electrode carefully after each experiment to prevent the electrode surface from absorbing other substances which will cover the electrode surface, reduce the contact area and affect the experimental data with errors. The interface of each electrode shall avoid mutual contact to avoid short circuits.

4.4 According to the same experimental method mentioned above, the 5 times diluent of pediatric Tylenol syrup (diluted with Mcllvaine buffer of pH=2.2) is determined, and the oxidation peak is recorded. Check the standard curve, calculate the concentration of APAP in the sample, and compare it with the mark value of APAP in the drug packaging box.

4.5 The 3mmol/L APAP solution with pH=2.2, pH=6.0 Mcllvaine buffer and 1.8mol/L sulfuric acid solution are scanned by cyclic voltammetry at the rates of 40mV/s and 250mV/s, respectively. CV curves are recorded and the oxidation and reduction peaks are observed at different pH and different scanning rates. The electrochemical oxidation mechanism of APAP proposed in the experimental principle is confirmed.

4.6 Experimental data processing requirements

4.6.1 Obtain the curve relationship between concentration and oxidation peak current according to the cyclic voltammograms of acetaminophen solutions with different concentrations (0.1mmol/L, 0.4mmol/L, 1mmol/L, 3mmol/L, 5mmol/L) at pH=2.2.

4.6.2 According to the size of the oxidation peak in the cyclic voltammetry of the sample, check the standard curve to calculate the concentration of APAP in the sample and compare it with the marked value of APAP in the drug packaging box.

4.6.3 According to the cyclic voltammograms of acetaminophen at different pH (pH=2.2, pH=6.0, 1.8mol/L sulfuric acid) solutions, discuss the relationship between the current potential value of oxidation and reduction peaks and the solution pH. Confirm the electrochemical oxidation mechanism of APAP proposed in the experimental principle.

5. Experimental HSE

Sulfuric acid has strong stimulation and corrosion on the skin, mucosa, and other tissues. Pay attention to protection when using it.

6. Questions

6.1 By comparing the scanning spectra in pH=2.2 and the strong acid medium, a new reduction peak appears in the strong acid. Please explain.

6.2 According to the ΔE_p at different scanning rates, is the electrochemical reaction of APAP on the electrode surface a reversible process in the system of pH=6.0?

6.3 At the same scanning rate in the medium of pH=6.0, pH=2.2 McIlvaine buffer, and strong acid, why does the oxidation potential value of APAP move with the change of acidity?

实验十六 ZnS∶Mn 纳米荧光材料的合成及其光学性质研究

一、实验目的

1. 掌握 ZnS∶Mn 纳米荧光材料的化学合成方法。
2. 了解纳米荧光材料的性质和用途。

二、实验原理

发光是物体内部以某种方式吸收能量，然后转化为光辐射的过程。发光技术近年来得到了广泛的应用。以固体发光为例，它可用于各种形式的光源、显示和显像技术、光电子器件、辐射场的探测及辐射剂量的记录等。

在过去十多年里，半导体纳米材料以其独有的特性，如大的比表面积、量子限域和其他效应，受到材料科学家的广泛关注。ZnS 是性能优良的Ⅱ～Ⅵ族半导体材料，禁带宽度为 3.66eV，在可见光范围内具有高的折射率和透射率，被广泛用于薄膜电光器件、红外线窗口、感应器以及激光器等。因为具有红外透明、荧光、磷光等特性，当硫化锌纳米粒子半径小于激子玻尔半径时，表现出量子限域效应，产生吸收光谱蓝移和高效荧光特性，可用于制备新一代高性能的光电传感器和平面显示器。

ZnS 作为一种性能独特的发光材料基质，通过改变其中的掺杂剂，可得到不同波段的高效可见辐射。在 ZnS 基质中掺入 Mn^{2+} 后，其发光量子效率大大提高。

本实验介绍一种低成本、易于操作的 ZnS∶Mn 荧光纳米材料的化学合成方法。其合成原理如下所示：

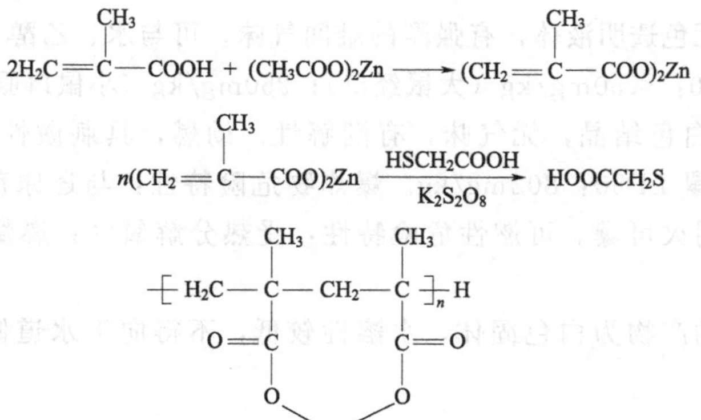

反应主要包括两步：(1) 通过调节单体甲基丙烯酸与链转移剂巯基乙酸的摩尔比，合成不同链长的聚甲基丙烯酸锌；(2) 聚甲基丙烯酸锌与 Na_2S 反应生成 ZnS∶Mn 纳米粒子。在纳米

粒子的形成过程中，聚甲基丙烯酸包覆了生成的 ZnS:Mn 纳米微粒，从而保证了纳米粒子的热、机械、化学稳定性和高分散性，其荧光效率也有明显提高。

三、仪器与试剂

1. 仪器

F4500 荧光光谱仪，X 射线衍射仪，透射电镜，球形冷凝管，三口瓶，水浴锅，烧杯，注射器（0.5mL、10mL、50mL 各 1 支）。

2. 试剂

单体甲基丙烯酸用常规方法除去阻聚剂。$Zn(CH_3COO)_2 \cdot 2H_2O$，$MnSO_4 \cdot H_2O$，$Na_2S \cdot 9H_2O$，$K_2S_2O_8$，$HSCH_2COOH$ 等。

四、实验步骤

1. ZnS:Mn 纳米材料的合成

将100mL 水、0.03mol 的乙酸锌和 5.5mL 的甲基丙烯酸加到带有冷凝管的 250mL 三口烧瓶中，氮气保护下升温搅拌。当温度升至 60℃时，分别向反应瓶中加入 0.06g 过硫酸钾和 0.03mL 的巯基乙酸（甲基丙烯酸与巯基乙酸的摩尔比约为 200∶1），在 80℃条件下反应 1.5h。待温度冷却至 50℃，向反应物中加入 20mL 0.06mol/L 的 $MnSO_4$ 水溶液。然后在 30min 内滴加 40mL 0.8mol/L 的 Na_2S 水溶液，滴加完毕后，使反应混合物在 80℃继续反应 30min。反应结束后取一滴反应液加去离子水稀释用于透射电镜测试。室温下将反应混合物进行离心分离，用甲醇洗涤产物，真空干燥，即可得到白色的聚甲基丙烯酸包覆的 ZnS:Mn 荧光产物。

2. 样品表征

ZnS:Mn 样品的发射光谱由 F4500 荧光光谱仪在室温下测定，激发波长为 332nm。进行荧光测试时，样品的浓度不宜过高。样品的荧光强度可以通过调节荧光光谱仪的狭缝宽度进行优化，以荧光强度不超过检测限为宜。X 射线粉末衍射在 Rigaku RU200B X 射线衍射仪上进行。粒子的形貌和聚集状态在 JEOL2010 透射电镜上进行，加速电压为 200kV。

五、实验 HSE

1. 巯基乙酸：无色透明液体，有强烈的难闻气味。可与水、乙醇、乙醚和普通溶剂混溶。急性毒性，LD50：＜50mg/kg（大鼠经口）；250mg/kg（小鼠口服）。

2. 过硫酸钾：白色结晶，无气味，有潮解性。助燃，具刺激性。毒性分级：中毒。急性毒性，口服-大鼠 LD50：802mg/kg。爆炸物危险特性：与还原剂、硫、磷等混合可爆；受热、撞击、明火可爆。可燃性危险特性：受热分解氧气；燃烧产生有毒氮氧化物烟雾。

3. 该实验产生的产物为白色固体，水溶性较低，不得向下水道倾倒，以免造成水管堵塞。

六、思考题

1. 如何解释添加少量 Mn^{2+} 会增强发光效率这一实验现象？

2. 甲基丙烯酸的作用是什么，它的用量如何确定？

Experiment 16　Synthesis and optical properties of ZnS∶Mn fluorescent nanomaterials

1. Objectives

1.1　Master the chemical synthesis method of ZnS∶Mn fluorescent nanomaterials.

1.2　Understand the properties and applications of fluorescent nanomaterials.

2. Principle

Luminescence is a process in which an object absorbs energy in some way and then converts it into light radiation. Luminescence technology has been widely used in recent years. For example, the luminescence of solid can be used for various forms of light source, display and development technology, optoelectronic devices, radiation field detection, and radiation dose recording.

In the past decade, semiconductor nanomaterials have received extensive attention from materials scientists for their unique properties, such as large specific surface area, quantum confinement, and other effects. ZnS is a kind of Ⅱ ~ Ⅵ semiconductor material with excellent properties. The bandgap is 3.66 eV and it has a high refractive index and transmission coefficient at the visible range. ZnS has been widely used in thin-film electro-optical devices, infrared windows, sensors, and lasers. Because of its infrared transparency, fluorescence, phosphorescence, and other characteristics, when the radius of ZnS nanoparticles is less than the exciton Bohr radius, it exhibits quantum confinement effect, a blue shift of absorption spectrum, and high-efficiency fluorescence characteristics, which can be used to prepare a new generation of high-performance photoelectric sensors and planar displays.

As a unique luminescent material matrix, ZnS can obtain high-efficiency visible radiation in different wavelength bands by changing the dopants. When the Mn^{2+} is incorporated into the ZnS matrix, the quantum efficiency of luminescence is greatly improved.

This experiment introduces a low-cost and easy-to-operate chemical synthesis method of ZnS∶Mn fluorescent nanomaterials. The principle of synthesis is as follows:

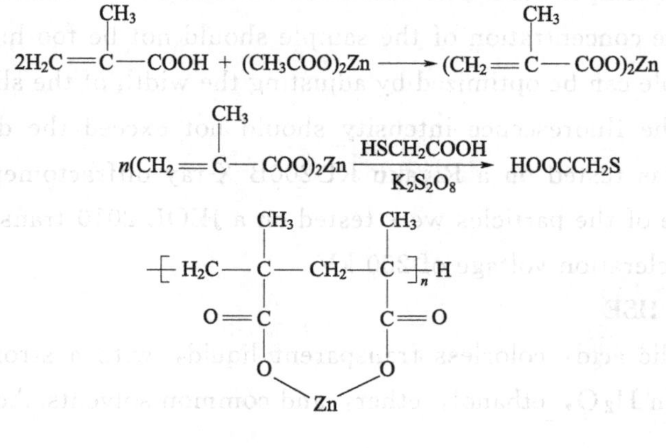

The reaction mainly consists of two steps: (1) Synthesis of zinc polymethacrylate with different chain lengths by adjusting the molar ratio of monomeric methacrylic acid to chain transfer agent thioglycolic acid; (2) Zinc polymethacrylate reacts with Na_2S to form ZnS: Mn nanoparticles. During the formation of nanoparticles, polymethacrylic acid encapsulates the generated ZnS: Mn nanoparticles, which ensures the thermal, mechanical, chemical stability, and high dispersion of the nanoparticles, and the fluorescence efficiency is also significantly improved.

3. Instruments and reagents

3.1 Instruments

F4500 fluorescence spectrometer, XRD, TEM, allihn condenser, three necks flask, water-bath, beakers, the injector (0.5mL, 10mL, and 50mL).

3.2 Reagents

Removal of inhibitors from monomeric methacrylic acid by conventional methods. $Zn(CH_3COO)_2 \cdot 2H_2O$, $MnSO_4 \cdot H_2O$, $Na_2S \cdot 9H_2O$, $K_2S_2O_8$, $HSCH_2COOH$, etc.

4. Procedures

4.1 Synthesis of ZnS:Mn Nanomaterials

100mL water, 0.03mol zinc acetate, and 5.5mL methacrylic acid were added to a 250mL three-necked flask with a condenser pipe. Heating and stirring under nitrogen protection. When the temperature was raised to 60℃, 0.06g potassium persulfate and 0.03mL thioglycolic acid (the molar ratio of methacrylic acid to thioglycolic acid was about 200 : 1) were added into the reaction flask, then the reaction was carried out at 80℃ for 1.5 h. After the temperature was cooled to 50℃, 20mL 0.06mol/L $MnSO_4$ was added into the reaction. Then, 40mL of 0.8mol/L Na_2S was added dropwise within 30 min. After that, the mixture was further reacted at 80℃ for 30 min. After the reaction, take one drop of the reaction solution and dilute with deionized water for transmission electron microscopy. The mixture was centrifuged at room temperature, and the product was washed with methanol and dried in vacuo to obtain a white polymethacrylic acid-coated ZnS:Mn fluorescent product.

4.2 Characterization

The emission spectrum of the ZnS:Mn sample was measured by the F4500 fluorescence spectrometer at room temperature. The excitation wavelength is 332nm. When performing a fluorescence test, the concentration of the sample should not be too high. The fluorescence intensity of the sample can be optimized by adjusting the width of the slit of the fluorescence spectrometer, and the fluorescence intensity should not exceed the detection limit. X-ray powder diffraction was tested on a Rigaku RU200B X-ray diffractometer. The morphology and aggregation state of the particles were tested on a JEOL 2010 transmission electron microscope with an acceleration voltage of 200 kV.

5. Experimental HSE

5.1 Thioglycolic acid: colorless transparent liquid, with a strong unpleasant odor. Solubility: miscible in H_2O, ethanol, ether, and common solvents. Acute toxicity, LD50:

<50mg/kg (rat oral); 250mg/kg (mice oral).

5.2 Potassium persulfate: white crystal, odorless, deliquescent. Combustion support, irritating. Toxicity classification: poisoning. Acute toxicity, Oral-rat LD50: 802mg/kg. Dangerous characteristics of explosives: mixed with reducing agents, sulfur, phosphorus, etc. can be explosive; heat, impact, open flames can be explosive. Flammability hazard characteristics: heat decomposes oxygen; burning produces toxic nitrogen oxide fumes.

5.3 The product in this experiment is a white solid with low water solubility. Do not pour it into the sewer to avoid blockage of the water pipe.

6. Questions

6.1 How to explain the experimental phenomenon that adding a small amount of Mn^{2+} will enhance fluorescence efficiency?

6.2 What is the function of methacrylic acid and how to determine its usage amount?

实验十七　8-羟基喹啉铝的合成和荧光性能表征

一、实验目的

1. 了解8-羟基喹啉铝的性质及应用。
2. 掌握8-羟基喹啉铝的制备原理和方法。
3. 掌握8-羟基喹啉铝的表征方法。
4. 了解分子荧光光谱仪的性能与结构，学会绘制激光光谱和荧光谱图。

二、实验原理

1. 8-羟基喹啉铝的基本性质

目前的绿光有机电致发光材料中，人们研究得最多的为8-羟基喹啉铝（AlQ_3）。1987年美国Kodak公司的C. W. Tang及其合作者报道了一种以8-羟基喹啉铝制成的发绿光的双层有机电致发光器件，人们开始对AlQ_3进行了广泛深入的研究。8-羟基喹啉铝（AlQ_3）作为一种较理想的有机电致发光材料，具有良好的成膜性和热稳定性（T_g=175℃）、独特的发光特性和电子传输特性［$10^{-4}cm^2/(V \cdot s)$］，被广泛应用于各种不同类型的有机电致发光器件。作为发光材料使用的8-羟基喹啉铝纯度须在95%以上。目前，实验室和工业上很难直接合成纯度在95%以上的8-羟基喹啉铝，需要通过复杂的色谱或升华提纯方法进一步提纯。若通过调整反应酸度和反应时间，可以直接合成纯度在95%以上的8-基喹啉铝，这种方法不仅能在实验室使用，也适合工业上大批量生产。8-羟基喹啉铝是黄色粉末，主要用于发光材料及电子传输材料，可用于聚氨酯塑料橡胶、皮革、纸张、纺织、涂料、木材等，也可用于农药、医药、合成金属缓蚀剂等方面。

化学反应和副反应方程式如图1所示，当pH=5.0时，开始有较多的黄绿色沉淀产生，产率也迅速增加，但是pH>6.5时，产物的纯度会明显降低。而当pH>7.0时，实验发现有白色絮状$Al(OH)_3$沉淀产生，体系中的酸度降低时，产物的纯度也随之降低，pH<6.5

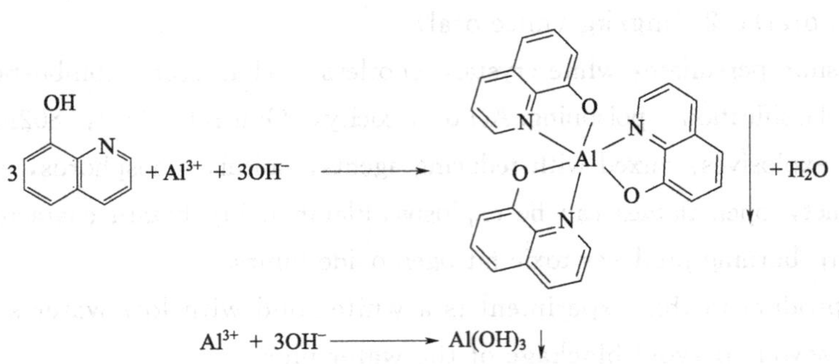

图 1　8-羟基喹啉铝的合成示意图

时体系的酸度对产物的纯度影响很小。实验中选择最佳 pH 值约为 6.5。

2. 8-羟基喹啉铝的制备方法

常规制备 8-羟基喹啉铝的方法反应时间长、产率低，而且要用真空蒸发法进行提纯，并且用到遇水即爆炸分解的氢化锂铝，增加了实验成本，并给实验带来了一定的复杂性和难度。本实验直接采用 8-羟基喹啉作为原料制备高纯 8-羟基喹啉铝的方法，保证产物纯度大于 95%。制备过程以 $Al_2(SO_4)_3$ 和 8-羟基喹啉为原料，通过反应制得草绿色固体，再进一步反复洗涤和烘干得到样品。该实验原料的纯度较高，整个过程不引入杂质离子，且产品质量稳定。

3. 荧光光谱性能的测定

荧光光谱可用于物质的定性及定量，由于物质结构不同，所吸收的紫外-可见光波长不同，所发射荧光波长也不同，利用这个光谱性质和标准图谱比对可鉴别物质。在一定频率和一定强度的激发光照射下，荧光物质（稀溶液体系）所产生的荧光强度与浓度呈线性关系，可进行定量分析。荧光光谱仪主要由四个部分组成：激发光源、样品池、双单色器系统和检测器。特殊点：有两个单色器，光源与检测器通常成直角。

三、仪器与试剂

1. 试剂

$Al_2(SO_4)_3$，8-羟基喹啉，乙醇，蒸馏水，$CDCl_3$，氨水。

2. 仪器

烧瓶，冷凝管，磁力加热搅拌机，容量瓶，抽滤瓶，布氏漏斗，电子天平，恒压漏斗，核磁管，核磁共振波谱仪，荧光光谱仪（日立），比色皿。

四、实验步骤

1. 8-羟基喹啉铝的制备

(1) 用天平分别称取 2.0g $Al_2(SO_4)_3 \cdot 18H_2O$ 和 1.5g 8-羟基喹啉，用量筒分别量取 20mL 乙醇和 5.0mL 水。将 8-羟基喹啉和 20mL 乙醇混合，将 $Al_2(SO_4)_3 \cdot 18H_2O$ 与 5.0mL 水混合。

(2) 将 8-羟基喹啉乙醇溶液倒入圆底烧瓶中，搅拌并控温在 60℃充分溶解。

(3) 将 $Al_2(SO_4)_3 \cdot 18H_2O$ 水溶液滴入圆底烧瓶中，并用 1∶10 的氨水调节 pH 值到黄绿色沉淀不再产生为止。

(4) 用冷凝管回流，在70℃时反应30min后，冷却至室温，进行抽滤，得到草绿色沉淀产物，用少量乙醇洗涤3次，得到纯度较高的产物。放入烘箱干燥1h后称重。

2. 8-羟基喹啉铝的表征

核磁共振谱氢谱（^1H NMR）：为了分析8-羟基喹啉铝中氢原子的位置，用核磁共振波谱仪在400MHz的频率下，溶剂为氘代氯仿（$CDCl_3$），对样品进行了测试，得到8-羟基喹啉铝的核磁共振谱氢谱。

3. 8-羟基喹啉铝的荧光性能

(1) 溶液的配制

取3只25mL的容量瓶，分别加入20.0μg/mL 8-羟基喹啉铝溶液0.00mL、1.00mL和5.00mL，稀释至刻度，摇匀，分别记为标0、标1、标2，其中，标0做空白试剂。

(2) 仪器操作步骤

①检查仪器所有电源确保开关处于关闭状态，然后通电；②打开主机开关、电脑开关等；③通过电脑进入分析程序。

(3) 绘制AlQ_3荧光发射谱图（对比不同激发波长下的最大发射波长λ_{EX}）

① 仪器正常后设定实验参数，更改实验名称等。

② 进行预扫后，开始对标液和样品扫描，画出曲线。

a. 将λ_{EX}固定在350nm，选择合适的实验条件，在400～650nm范围内扫描即得荧光发射光谱，从谱图找出最大λ_{EM}值。

b. 将λ_{EX}固定在420nm，选择合适的实验条件，在400～650nm范围内扫描即得荧光发射光谱，从谱图找出最大λ_{EM}值。

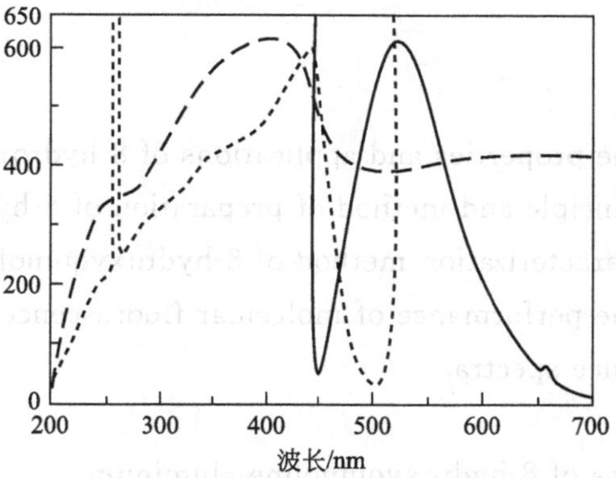

图2 制备的AlQ_3的荧光光谱

用荧光光谱仪测定样品的荧光谱图。

图2是激发波长为420nm时的AlQ_3的荧光光谱图，可以看出其发射峰位于510nm处，绿光半峰宽大约为83nm，且没有其他杂质峰，这进一步证明了所制备的物质为AlQ_3，与文献报道一致。

(4) 扫描完数据存盘后上传至网盘。

(5) 退出分析程序，关机，关电源，取出比色皿，洗净、保存所有玻璃器皿，实验结束。

(6) 实验报告根据网盘中下载的数据撰写。

五、实验 HSE

1. $Al_2(SO_4)_3 \cdot 18H_2O$：无毒，其粉尘对眼睛、黏膜有一定的刺激作用，称量样品时务必在通风橱中进行操作。

2. 8-羟基喹啉：易溶于乙醇、丙酮、氯仿、苯和硫酸，不溶于水和乙醚，但能溶于强酸或强碱，在碱中电离成负离子，在酸中能结合氢离子。低毒，LD50：4800mg/kg，腐蚀性较小。

3. 实验中需使用烘箱等高温设备，必须做好防护，防止高温烫伤，全程不得离开实验室。

4. 使用氨水时要注意保护好口鼻和呼吸道，使用时要戴上手套。

六、思考题

1. 有没有其他制备 8-羟基喹啉铝的方法？
2. 为什么测试 ^1H NMR 时用氘代氯仿（$CDCl_3$）作为溶剂？
3. 荧光光谱仪为什么要设两个单色器？
4. 荧光光谱仪和紫外-可见分光光度计有哪些不同点？

Experiment 17 Synthesis and fluorescence characterization of 8-hydroxyquinoline aluminum

1. Objectives

1.1 Understand the properties and applications of 8-hydroxyquinoline aluminum.

1.2 Master the principle and method of preparation of 8-hydroxyquinoline aluminum.

1.3 Master the characterization method of 8-hydroxyquinoline aluminum.

1.4 Understand the performance of molecular fluorescence spectrometers and learn to draw laser and fluorescence spectra.

2. Principle

2.1 Basic properties of 8-hydroxyquinoline aluminum

Among the organic electroluminescent materials that are currently emitting green light, the most studied material is 8-hydroxyquinoline aluminum (AlQ_3). The 8-hydroxyquinoline aluminum not only has good electron transportability [10^{-4} $cm^2/(V \cdot s)$] and high photoluminescence efficiency but also has a relatively high glass transition temperature ($T_g = 175℃$). The 8-hydroxyquinoline aluminum has a good film formation type during vapor deposition, and the stability of the device is good. Therefore, a large number of researches are studied on 8-hydroxyquinoline aluminum and its derivatives, including its luminous efficiency, device aging, luminescence intensity and fluorescence lifetime. Its excellent luminescent properties and physical properties have been used as star molecules in organic electroluminescent materials.

2.2 Preparation method of 8-hydroxyquinoline aluminum

Main raw materials (Information and Properties)

$Al_2(SO_4)_3 \cdot 18H_2O$: The molecular weight is 666.43, which is soluble in water. Aluminum sulfate cannot be dissolved in pure sulfuric acid (only coexisting). It is dissolved in water together with sulfuric acid in sulfuric acid solution. Therefore, the solubility of aluminum sulfate in sulfuric acid is the solubility of aluminum sulfate in water. The precipitation at room temperature contains 18 molecules of crystal water. It is not easy to weather and lose crystal water. It is relatively stable, heating will lose water, and high temperature will decompose into alumina and sulfur oxides. Heating to 770℃ began to decompose into alumina, sulfur trioxide, sulfur dioxide, and water vapor, soluble in water, acid, and alkali, insoluble in ethanol, hydrolyzed to form aluminum hydroxide. The aqueous solution boils for a long time to form basic aluminum sulfate. Industrial products are gray-white flakes, granules, or lumps. They are light green due to low iron salts and yellowish due to oxidation of low-cost iron salts. The crude product is a grayish-white fine crystal structure porous. Non-toxic, dust can irritate the eyes.

8-hydroxyquinoline aluminum: White or light yellow crystal or crystalline powder; insoluble in water and ether, soluble in ethanol, acetone, chloroform, benzene or dilute acid; less corrosive, can sublimate. Low toxicity, melting point: 75～76℃, boiling point: 267℃, density: 1.03, soluble in ethanol, acetone, chloroform, benzene, and sulfuric acid, almost insoluble in water. 8-hydroxyquinoline is amphoteric, soluble in strong acids and strong bases, ionized into negative ions in alkali, and capable of binding hydrogen ions in acid, and has the lowest solubility at pH=7.

The conventional method for preparing 8-hydroxyquinoline aluminum has a long reaction time and low yield, and is purified by vacuum evaporation, and uses lithium aluminum hydride which decomposes in the presence of water, which increases the experimental cost and brings the experiment. The main goal of this experiment was to develop a new and simple method for the preparation of 8-hydroxyquinoline aluminum to test the fluorescence spectrum properties.

The preparation reaction process uses $Al_2(SO_4)_3$ and 8-hydroxyquinoline as raw materials to prepare a grass green sample. Further washing with water, centrifugation, and drying is obtained. The reaction mechanism equation is:

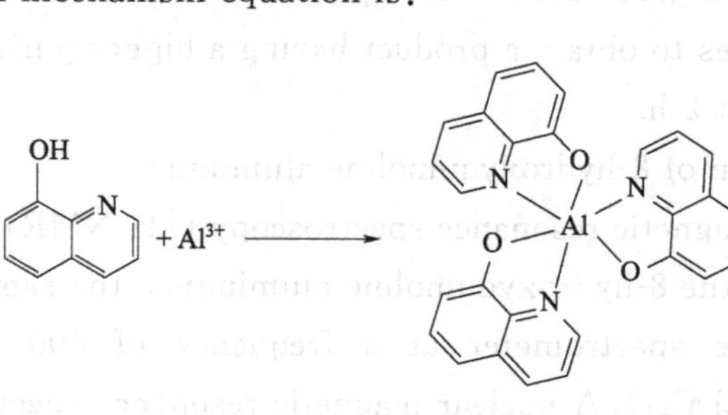

Fig. 1 Synthesis of 8-hydroxyquinoline aluminum

The purity of 8-hydroxyquinoline aluminum in the experimental step is high, the impurity ions are not introduced in the whole process, and the product quality is stable.

2.3 Fluorescence spectral properties

Fluorescence spectroscopy can be applied to the qualitative and quantitative determination of substances. Due to the different structure of the material, the wavelength of the ultraviolet-visible light absorbed is different, and the wavelength of the emitted fluorescent light is also different. This property can be used to identify the substance. Under the excitation of a certain frequency and a certain intensity of excitation light, the fluorescence intensity produced by the fluorescent substance (dilute solution system) has a linear relationship with the concentration, which can be quantitatively analyzed. The fluorescence spectrometer consists of four main components: the excitation source, the sample cell, the dual monochromator system, and the detector. Special point: There are two monochromators. The light source and the detector are usually at right angles.

3. Instruments and Reagents

3.1 Instruments

Flask, condenser, magnetic heating mixer, volumetric flask, Suction filter bottle, büchner funnel, analytical balance, constant pressure funnel, nuclear magnetic tube, nuclear magnetic resonance spectrometer, fluorescence spectrometer (Hitachi), Cuvette.

3.2 Reagents

$Al_2(SO_4)_3$, 8-hydroxyquinoline, ethanol, distilled water, $CDCl_3$ ammonium hydroxide.

4. Procedures

4.1 Preparation of 8-hydroxyquinoline aluminum

(1) Using analytical balance to weigh 2.0g $Al_2(SO_4)_3 \cdot 18H_2O$ and 1.5g 8-hydroxyquinoline, respectively. The 8-hydroxyquinoline was mixed with 20mL ethanol, and $Al_2(SO_4)_3 \cdot 18H_2O$ was mixed with 5.0mL water.

(2) The 8-hydroxyquinoline ethanol solution was poured into a round bottom flask, magnetically stirred, and the temperature was controlled to fully dissolve at 60℃.

(3) An aqueous solution of $Al_2(SO_4)_3 \cdot 18H_2O$ was dropped into a round bottom flask, and pH was adjusted to a yellow-green suspension with (1∶10) ammonia water.

(4) The mixture was refluxed with a condenser pipe and reacted at 70℃ for 3 h.

(5) The reactant was cooled to room temperature for about 0.5h, and suction filtration was carried out to obtain a green precipitated product, which was washed with ethanol and deionized water for 5 times to obtain a product having a higher purity. The product is placed in a drying oven for about 2 h.

4.2 Characterization of 8-hydroxyquinoline aluminum

Hydrogen nuclear magnetic resonance spectroscopy (^1H NMR). To analyze the position of the hydrogen atom in the 8-hydroxyquinoline aluminum, the sample was tested by a nuclear magnetic resonance spectrometer at a frequency of 400 MHz. The solvent was deuterated chloroform ($CDCl_3$). A nuclear magnetic resonance spectrum of 8-hydroxyquinoline aluminum will be obtained.

4.3 Fluorescence properties of 8-hydroxyquinoline aluminum

(1) Solution preparation

Take three 25mL volumetric flasks, and add 20. 0 μg/mL 1,8-hydroxyquinoline aluminum solution 0. 00mL, 1. 00mL, and 5. 00mL, dilute to the mark and shake. The labels are 0, 1, and 2, and 0 is the blank reagent.

(2) Instrument operation steps

Check that all power switches in the instrument are off and then powered on; Turn on the main switch, computer switch, etc. ; Enter the analysis program through the computer.

(3) Draw fluorescence emission spectra of AlQ_3: (Compare the maximum emission wavelength λ_{EX} under different excitation wavelengths)

Set the test parameters after the instrument is normal; change the name of the experiment, etc.

After the pre-sweep, start scanning the standard solution and sample and draw the curve.

A) Fix λ_{EX} at 350nm, select appropriate experimental conditions, scan the fluorescence emission spectrum in the range of 400～650nm, and find the maximum λ_{EM} value from the spectrum.

B) Fix λ_{EX} at 420nm, select the appropriate experimental conditions, scan the fluorescence emission spectrum in the range of 400～650nm, and find the maximum λ_{EM} value from the spectrum.

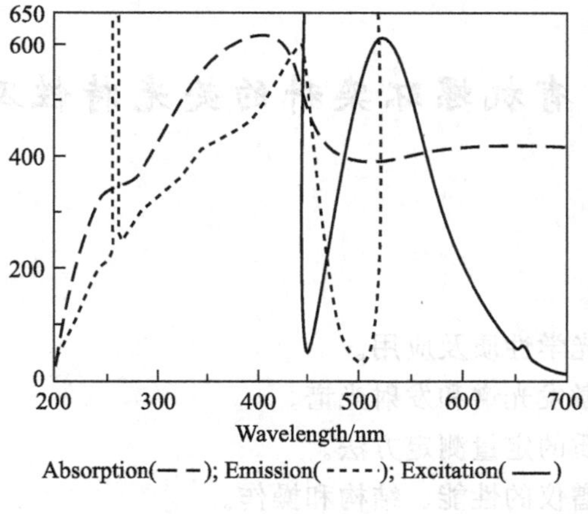

Fig. 2 Fluorescence spectra of prepared AlQ_3

The PL spectrum of the sample was measured by a fluorescence spectrometer.

Fig. 2 shows the fluorescence spectrum of AlQ_3 at an excitation wavelength of 420nm. It can be seen that the emission peak at 510nm has a green half-width of about 83nm and no other impurity peaks. This further proves that the prepared material is AlQ_3 (Consistent with the literature) .

(4) After scanning the data, upload it to the network disk;

(5) Exit the analysis program, shut down, turn off the power, remove the cuvette, wash and save all the glassware, and the experiment is over.

(6) The experimental data is exported by downloading data from the network disk.

5. Experimental HSE

5.1 $Al_2(SO_4)_3 \cdot 18H_2O$: Non-toxic. Aluminum sulfate dust has an irritating effect on the eyes and mucous membranes. When weighing the sample, it must be operated in the fume hood.

5.2 8-hydroxyquinoline aluminum: insoluble in water and ether, soluble in ethanol, acetone, chloroform, benzene or dilute acid; less corrosive, can sublimate. Low toxicity. LD50: 4800mg/kg. Less corrosive.

5.3 The high-temperature equipment of the oven is used in the experiment, and the protection shall be well done to prevent high-temperature scald, and the operator shall not leave the laboratory through the whole process.

5.4 Wear gloves when using ammonia, and pay attention to protect the mouth, nose, and respiratory tract.

6. Questions

6.1 Is there any other method for preparing 8-hydroxyquinoline aluminum?

6.2 Why does deuterated chloroform ($CDCl_3$) used as a solvent in the ^1H NMR tested?

6.3 Why do two monochromators in the fluorescence spectrophotometer?

6.4 What are the differences between the fluorescence spectrophotometer and the UV-vis spectrophotometer?

实验十八　有机螺环染料的荧光特性及其定量分析

一、实验目的

1. 了解螺环染料的光学性质及应用。
2. 测量螺环染料的激发光谱和发射光谱。
3. 掌握通用荧光物质的定量测定方法。
4. 熟悉分子荧光光谱仪的性能、结构和操作。

二、实验原理

1. 荧光的产生过程

物质分子或原子在一定条件下吸收辐射能而被激发到较高电子能态后，在返回基态的过程中将以不同的方式释放能量。如在分子吸收分光光度法中，受激分子以热能的形式释放多余的能量，测量的是物质对辐射的吸收，属吸收光谱法；而发光分析是受激分子或原子以发射辐射的形式释放能量，测量的是物质分子或原子自身发射辐射的强度，属发射光谱法。

激发光谱曲线：固定测量波长（选最大发射波长），化合物发射的荧光强度与激发光波长的关系曲线。

荧光（发射）光谱曲线：固定激发光波长（选最大激发波长），化合物发射的荧光强度与发射光波长关系曲线。

电子跃迁到不同激发态能级，吸收不同波长的能量，产生不同吸收带，但均回到第一激发单重态的最低振动能级，再跃迁回到基态，产生波长一定的荧光，所以发射光谱的形状与激发波长无关。

2. 荧光的产生与分子结构的关系

(1) 跃迁类型：$\pi^* \rightarrow \pi$ 的荧光效率高，系间跨越过程的速率常数小，有利于荧光的产生。

(2) 共轭效应：提高共轭度有利于增加荧光效率并产生红移。

(3) 刚性平面结构：可降低分子振动，减少与溶剂的相互作用，故具有很强的荧光。如荧光素和酚酞有相似结构，荧光素有很强的荧光，酚酞却没有。

(4) 取代基效应：芳环上有供电基，使荧光增强。

3. 有机螺环染料的荧光特性

有机螺环类染料为一种价廉易得的荧光染料，在水相中能稳定产生极强的荧光，且具有较大的摩尔吸光系数及稳定的吸收波长和荧光发射波长。在碱性条件下，有机螺环类染料能形成具有内酰胺的螺环结构（图 1），为无吸收且无荧光类化合物。而在酸性条件下，内酰亚胺化合物形成一种在长波段有强吸收和强红色荧光的化合物。近年来，该类化合物被逐渐应用于分子识别领域。当检测物与其发生识别反应之后螺环打开，恢复其共轭大 π 键结构，从而导致检测体系的吸收及荧光信号均显著升高。目前，利用此机理来检测各种活性生物信号分子的荧光探针已被大量开发。

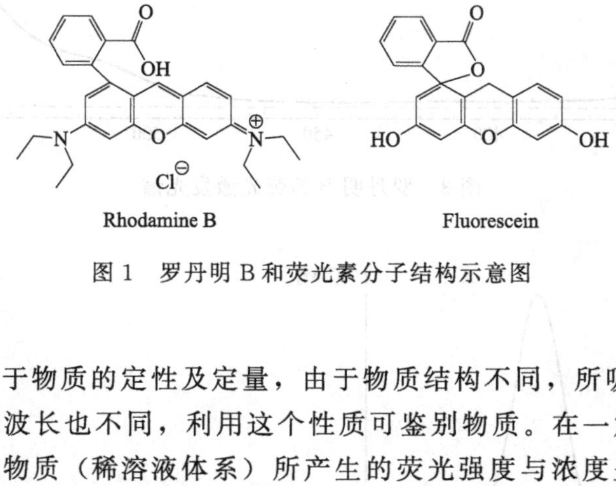

图 1　罗丹明 B 和荧光素分子结构示意图

4. 荧光分析

荧光分析可应用于物质的定性及定量，由于物质结构不同，所吸收的紫外-可见光波长不同，所发射荧光波长也不同，利用这个性质可鉴别物质。在一定频率和一定强度的激发光照射下，荧光物质（稀溶液体系）所产生的荧光强度与浓度呈线性关系，可进行定量分析。荧光分析法具备操作简单、选择性好、灵敏度高等优点而受到科学工作者越来越多的关注。

三、仪器与试剂

1. 仪器

移液枪（1000μL），移液管（50mL），50mL 容量瓶，电子天平，荧光光谱仪，比色皿。

2. 试剂

罗丹明B，荧光素，乙醇，蒸馏水，罗丹明B储备液（5μg/mL），荧光素储备液（5μg/mL），罗丹明 B 未知溶液，荧光素未知溶液。

四、实验步骤

1. 溶液的配制

（1）标准溶液的配制：按要求分别配制罗丹明 B 和荧光素的系列标准溶液各 4 个。

（2）选取其中一个溶液，分别测绘罗丹明 B 和荧光素的激发光谱和荧光光谱，并找到它们各自的最大激发波长和最大发射波长。

（3）分别以罗丹明 B 和荧光素的最大激发波长和最大发射波长，测定标准溶液的荧光强度，在相同条件下测定未知样品的荧光强度。

2. 荧光标准曲线绘制和未知溶液定量测量

（1）荧光仪的操作步骤参照实验十七。固定激发波长，扫描罗丹明 B 和荧光素发射波长得最大发射波长分别为 λ_{Em}＝572.5nm（图 2）和 λ_{Em}＝511.5nm（图 4）。

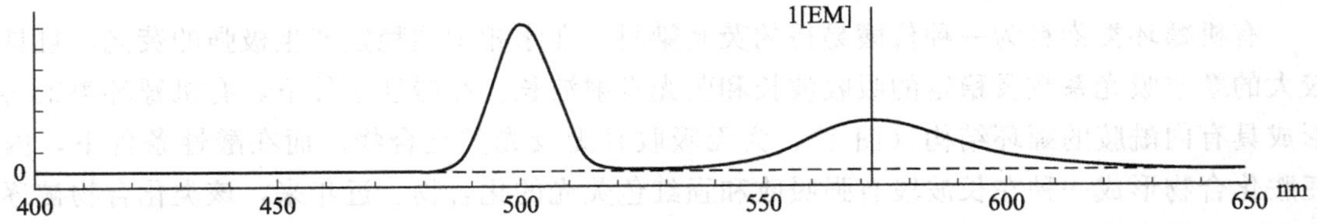

图 2　罗丹明 B 的荧光发射光谱

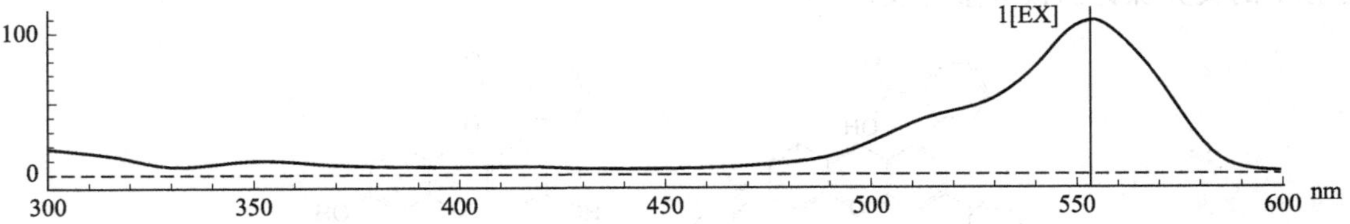

图 3　罗丹明 B 的荧光激发光谱

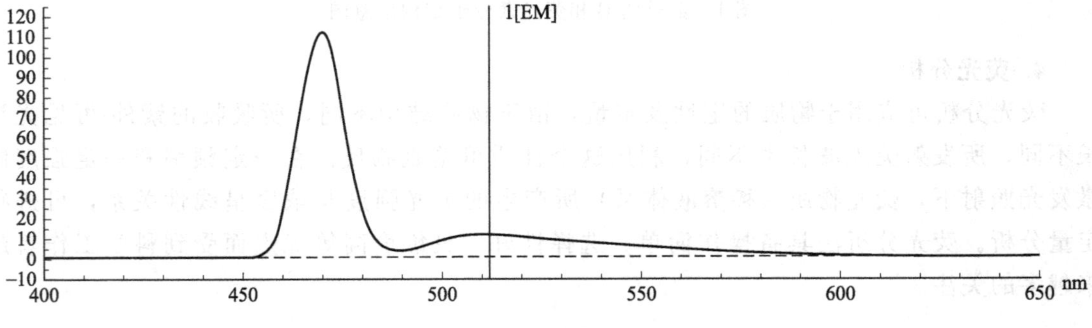

图 4　荧光素的荧光发射光谱

（2）再以 λ_{Em}＝572.5nm 和 λ_{Em}＝511.5nm 分别作为固定发射波长，扫描激发波长可以分别得到最大激发波长 λ_{Ex}＝553.5nm（图 3）和 λ_{Ex}＝490nm（图5）。

（3）分别在 λ_{Em}＝572.5nm 和 λ_{Em}＝511.5nm 条件下分别测量罗丹明 B 和荧光素的 4 个不同浓度（0.02μg/mL、0.04μg/mL、0.06μg/mL 和 0.08μg/mL）的标准样品信号强度，绘制荧光标准工作曲线，记入表 1 和表 2。

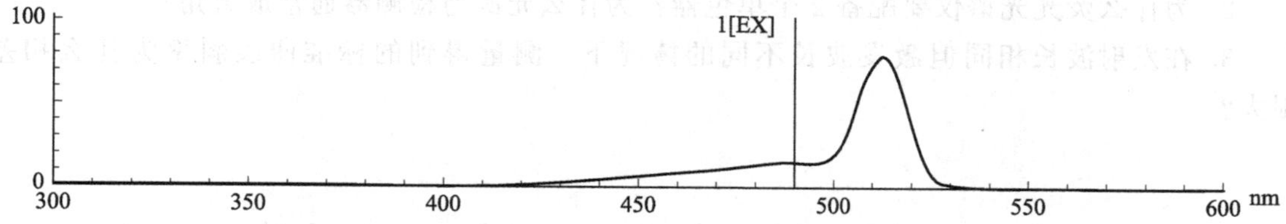

图 5　荧光素的荧光激发光谱

表 1　不同浓度罗丹明 B 标准样品的测量结果

样品编号	标样浓度/(μg/mL)	测量信号强度
1	0.02	
2	0.04	
3	0.06	
4	0.08	

表 2　不同浓度荧光素标准样品的测量结果

样品编号	标样浓度/(μg/mL)	测量信号强度
1	0.02	
2	0.04	
3	0.06	
4	0.08	

(4) 作线性回归线，计算相关系数 R，并测量未知罗丹明 B 和荧光素样品信号强度，代入工作曲线可求得未知样品的浓度。

五、实验 HSE

1. 罗丹明 B：又称玫瑰红 B，或碱性玫瑰精，俗称花粉红，是一种具有鲜桃红色的人工合成的染料。分子量 479.01，易溶于水、乙醇，微溶于丙酮、氯仿、盐酸和氢氧化钠溶液。水溶液为蓝红色，稀释后有强烈荧光，醇溶液为红色荧光。经老鼠试验发现，罗丹明 B 会引致皮下组织生肉瘤，被怀疑是致癌物质。罗丹明 B 在溶液中有强烈的荧光，用于实验室中细胞荧光染色剂、有色玻璃、特色烟花爆竹等。曾经用作食品添加剂，但后来实验证明罗丹明 B 会致癌，现在已不允许用作食品染色。2017 年 10 月 27 日，世界卫生组织国际癌症研究机构公布的致癌物清单初步整理参考，罗丹明 B 在Ⅲ类致癌物清单中。

2. 荧光素：又名荧光黄、荧光生、荧光红。有两种变体：稳定的红色变体 B 及黄色变体 A。分子量为 332.31。B 为红色带绿色荧光的结晶粉末，溶于热醇、热苯胺、热丙酮、热甲酸，微溶于水、醇、醚、乙酸，不溶于石油醚。A 为黄色无定形粉末，熔点 314～316℃，溶于丙酮、甲醇、甲酸，微溶于水、醇、醚、氯仿、苯、乙酸、二甲苯、硝基苯，不溶于石油醚，加热时变为红色变体，用作化学分析的指示剂、生物染色剂。由间苯二酚与邻苯二甲酸酐反应而得。

六、思考题

1. 在测量荧光光谱曲线时如何区分荧光发射峰和激发光散射峰？

2. 为什么荧光光谱仪要配备2个单色器？为什么光源与检测器通常成直角？

3. 在发射波长相同但激发波长不同的情况下，测量得到的标准曲线斜率为什么相差很大？

Experiment 18 Fluorescence characteristics and quantitative analysis of organic spiro dyes

1. Objectives

1.1 Understand the optical properties and applications of spiro dyes.

1.2 Measuring the excitation and emission spectra of spiro dyes.

1.3 Master the quantitative measurement method of universal fluorescent substances.

1.4 Understand the performance, structure, and operation of the molecular fluorescence spectrometer.

2. Principle

2.1 Fluorescence generation process

After a molecule or atom absorbs radiant energy under certain conditions and is excited to a higher electronic energy state, it will release energy in different ways during the process of returning to the ground state. For example, in molecular absorption spectrophotometry, stimulated molecules release excess energy in the form of thermal energy, which measures the absorption of radiation by substances, which belongs to absorption spectroscopy; while luminescence analysis is the release of stimulating molecules or atoms in the form of emitted radiation Energy, which measures the intensity of radiation emitted by a substance molecule or atom, belongs to emission spectrometry.

Excitation spectrum curve: fixed measurement wavelength (select the maximum emission wavelength), the relationship between the fluorescence intensity emitted by the compound and the excitation light wavelength.

Fluorescence (emission) spectral curve: fixed excitation light wavelength (select the maximum excitation wavelength), the relationship between the fluorescence intensity emitted by the compound and the wavelength of the emitted light.

The electron transitions to different excited-state energy levels absorb different wavelengths of energy, produces different absorption bands, but all return to the lowest vibrational energy level of the first excited singlet state and then transition back to the ground state, producing fluorescence at a certain wavelength, so the shape of the emission spectrum has nothing to do with the excitation wavelength.

2.2 The relationship between the generation of fluorescence and molecular structure

(1) Transition type: The high efficiency $\pi^* \rightarrow \pi$ fluorescence, and the small rate constant of the intersystem crossing process are conducive to the generation of fluorescence.

(2) Conjugation effect: Increasing the degree of conjugation is beneficial to increase the

fluorescence efficiency and produce a redshift.

(3) Rigid planar structure: Rigid planar structure can reduce molecular vibration and reduce interaction with solvent, so it has strong fluorescence. For example, fluorescein and phenolphthalein have similar structures, fluorescein has strong fluorescence, but phenolphthalein does not.

(4) Substituent effect: There is a donor group on the aromatic ring, which enhances the fluorescence.

2.3 Fluorescence properties of organic spiro dyes

Organic spiro dyes are inexpensive fluorescent dyes that can stably produce extremely strong fluorescence in the aqueous phase and have a large molar absorption coefficient and stable absorption wavelength and fluorescence emission wavelength. Under alkaline conditions, organic spiro dyes can form a spiro ring structure with lactams (Fig. 1), which are non-absorbing and non-fluorescent compounds. Under acidic conditions, the imide compound forms a compound with strong absorption and strong red fluorescence in the long-wavelength band. In recent years, these compounds have been gradually applied in the field of molecular recognition. When the detection substance and its recognition reaction occur, the spiro ring is opened, and its conjugated large π bond structure is restored, resulting in a significant increase in the absorption and fluorescence signal of the detection system. Currently, fluorescent probes that use this mechanism to detect various active biological signal molecules have been developed in large quantities.

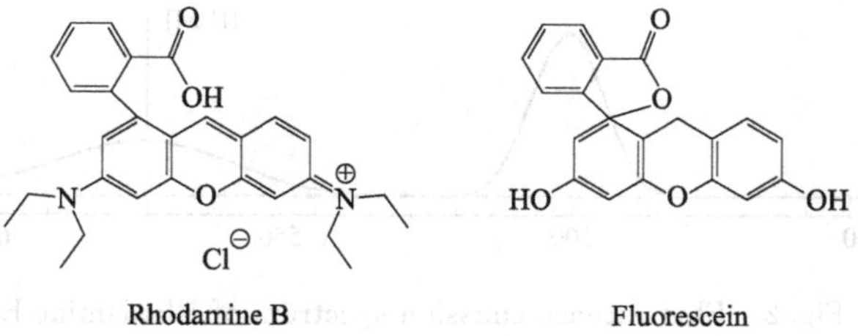

Fig. 1 Molecular structure of Rhodamine B and Fluorescein

2.4 Fluorescence analysis

Fluorescence analysis can be applied to the qualitative and quantitative determination of substances. Due to the different substance structures, the absorbed ultraviolet-visible wavelengths are different, and the emitted fluorescence wavelengths are also different. Using this property, substances can be identified. Under the excitation light of a certain frequency and a certain intensity, the fluorescence intensity (dilute solution system) generated by the fluorescent substance and the concentration are linearly related and can be quantitatively analyzed. The fluorescence analysis method has the advantages of simple operation, good selectivity, and high sensitivity, which has attracted more and more attention from scientists.

3. Instruments and Reagents

3.1 Instruments

Transfer liquid gun (1000μL), pipette (50mL), 50mL volumetric flask, electronic balance, fluorescence spectrometer, cuvette.

3.2 Reagents

Rhodamine B, fluorescein, ethanol, distilled water, rhodamine B stock solution (5μg/mL); fluorescein stock solution (5μg/mL), Rhodamine B unknown solution, luciferin unknown solution.

4. Procedure

4.1 Solution configuration

(1) Preparation of standard solution: Dispose of Rhodamine B and fluorescein series standard solutions respectively as required.

(2) Select one of the solutions and plot the excitation and fluorescence spectra of Rhodamine B and fluorescein respectively, and find their respective optimal excitation wavelength and strongest emission wavelength.

(3) Measure the fluorescence intensity of the standard solution with the maximum excitation wavelength and emission wavelength of Rhodamine B and fluorescein respectively, and measure the fluorescence intensity of the unknown sample under the same conditions.

4.2 Fluorescence standard curve drawing and quantitative measurement of an unknown solution

(1) Refer to Experiment 17 for the operation steps of the fluorometer. With a fixed excitation wavelength, scanning the emission wavelengths of Rhodamine B and fluorescein, the best emission wavelengths were λ_{Em}=572.5nm (Fig. 2) and λ_{Em}=511.5nm (Fig. 4).

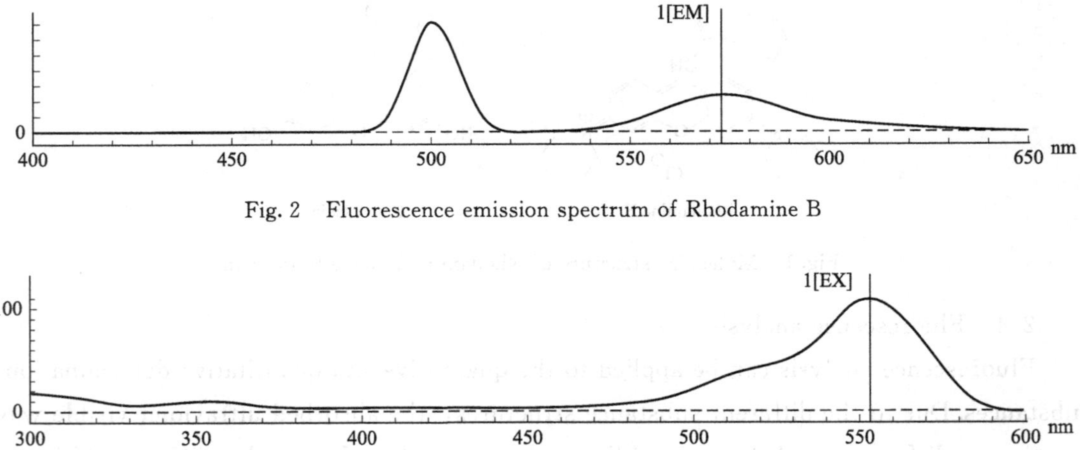

Fig. 2 Fluorescence emission spectrum of Rhodamine B

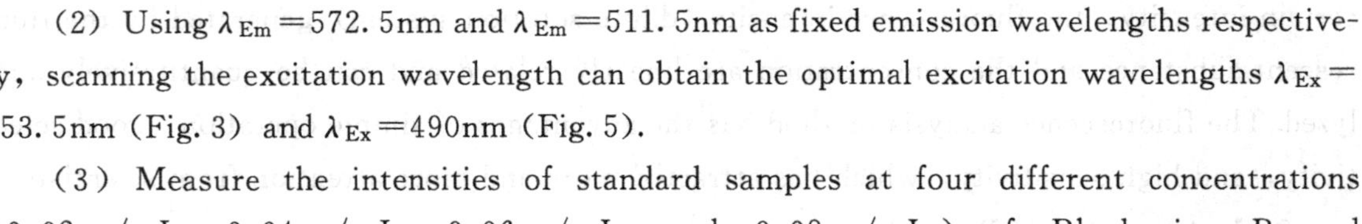

Fig. 3 Fluorescence excitation spectrum of Rhodamine B

(2) Using λ_{Em}=572.5nm and λ_{Em}=511.5nm as fixed emission wavelengths respectively, scanning the excitation wavelength can obtain the optimal excitation wavelengths λ_{Ex}=553.5nm (Fig. 3) and λ_{Ex}=490nm (Fig. 5).

(3) Measure the intensities of standard samples at four different concentrations (0.02μg/mL, 0.04μg/mL, 0.06μg/mL, and 0.08μg/mL) of Rhodamine B and fluorescein at λ_{Em} = 572.5nm and λ_{Em} = 511.5nm, respectively. Draw Standard working curve in table 1 and table 2.

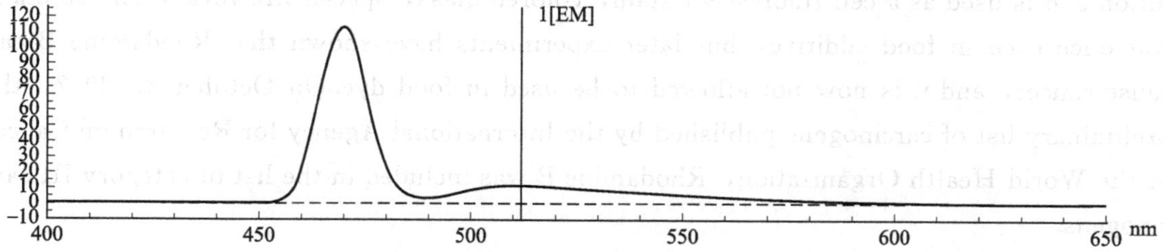

Fig. 4　Fluorescence emission spectrum of Fluorescein

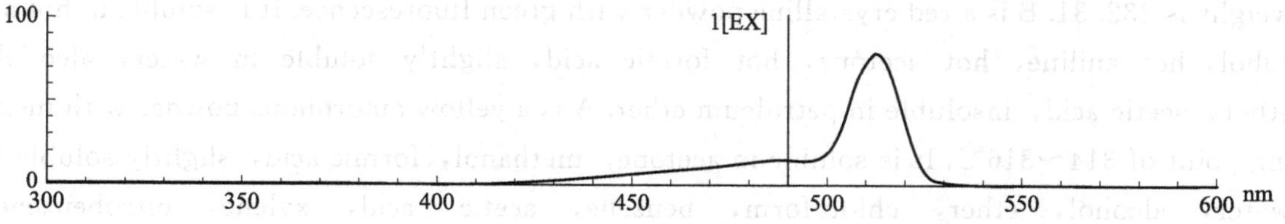

Fig. 5　Fluorescence excitation spectrum of Fluorescein

Table 1　Measurement results of different concentrations of rhodamine B standard samples

Sample number	Concentration/(μg/mL)	Signal intensity
1	0.02	
2	0.04	
3	0.06	
4	0.08	

Table 2　Measurement results of different concentrations of fluorescein standard samples

Sample number	Concentration/(μg/mL)	Signal intensity
1	0.02	
2	0.04	
3	0.06	
4	0.08	

(4) Make a linear regression line, calculate the correlation coefficient R, and measure the signal intensity of the unknown Rhodamine B and fluorescein samples, and bring it into the working curve to find the concentration of the unknown sample.

5. Experimental HSE

5.1　Rhodamine B: Also known as rose-red B, or basic rose essence, commonly known as flower pink, is a synthetic dye with a bright pink color. The molecular weight is 479.01. It is easily soluble in water and ethanol, and slightly soluble in acetone, chloroform, hydrochloric acid, and sodium hydroxide solution. The aqueous solution is blue-red, with strong fluorescence after dilution, and the alcohol solution is red fluorescence. Experiments in mice have revealed that Rhodamine B can cause subcutaneous tissue sarcoma, which is suspected to be a carcinogen. Rhodamine B has strong fluorescence in solution and is used as a cell fluorescent stain，colored glass，special fireworks，and so on. It was once used in food additive，but later experiments have shown that Rhodamine B can cause cancer，and it is now not allowed to be used in food dye. On October 27，2017，the preliminary list of carcinogens published by the International Agency for Research on Cancer of the World Health Organization，Rhodamine B was included in the list of category III carcinogens.

5.2 Fluorescein：also known as fluorescent yellow，fluorescent raw，and fluorescent red. There are two variants：the stable red variant B and the yellow variant A. The molecular weight is 332.31. B is a red crystalline powder with green fluorescence. It is soluble in hot alcohol，hot aniline，hot acetone，hot formic acid，slightly soluble in water，alcohol，ether，acetic acid，insoluble in petroleum ether. A is a yellow amorphous powder with melting point of 314～316℃. It is soluble in acetone，methanol，formic acid，slightly soluble in water，alcohol，ether，chloroform，benzene，acetic acid，xylene，nitrobenzene，insoluble in petroleum ether. It becomes a red variant when heated. It used as an indicator of chemical analysis，biological stain. It is derived from the reaction of resorcinol and phthalic anhydride.

6. Questions

6.1 How to distinguish fluorescence emission peak and excitation light scattering peak when measuring fluorescence spectrum curve?

6.2 Why is the fluorescence spectrometer equipped with 2 monochromators? Why is the light source and detector usually at right angles?

6.3 When the emission wavelength is the same but the excitation wavelength is different，why the slope of the measured standard curve differs greatly?

实验十九　离子液体 $bmimBF_4$ 的制备

一、实验目的

1. 掌握离子液体 $bmimBF_4$ 的制备方法。
2. 了解环境友好的离子液体的性质及应用。

二、实验原理

离子液体（Ionic liquids，ILs）又称室温熔融盐（Room temperature molten salt），在室温或室温附近（一般<100℃）呈现液体状态，主要由含氮、磷的有机阳离子和无机或有机阴离子所组成。通常离子化合物只有在高温下才能变成液体，而离子液体的阴阳离子体积大，结构松散，导致阴阳离子间的静电作用力较低，熔点接近室温。

离子液体与传统的有机溶剂相比，具有一系列突出的优点：较宽的液态范围、较低的蒸气压、不易燃、可调控的物化性能、较宽的电化学窗口、易回收利用等。离子液体作为有机合成的反应介质和催化剂、分离科学的溶剂和萃取剂、电化学的电解液等得到广泛应用，已成为绿色化学研究的重要组成部分。

1-丁基-3-甲基咪唑四氟硼酸盐（$bmimBF_4$）是最常见的室温离子液体。其合成路线如下：

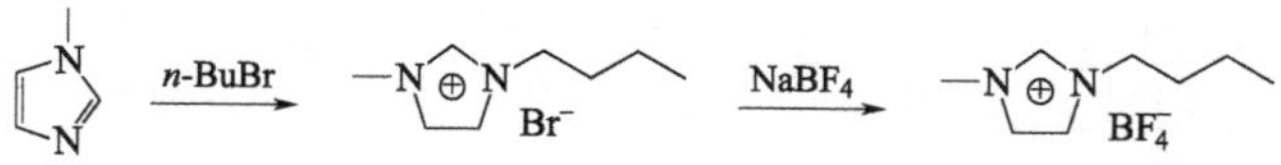

三、仪器与试剂

1. 仪器

冷凝管，恒压滴液漏斗，150mL 三口烧瓶，100mL 圆底烧瓶，温度计，布氏漏斗，抽滤瓶，玻璃棒，分液漏斗，电子天平，铁架台，加热套，变压器，旋转蒸发仪，循环水真空泵。

2. 试剂

N-甲基咪唑，溴代正丁烷，乙酸乙酯，甲醇，二氯甲烷，四氟硼酸钠（$NaBF_4$）。

四、实验步骤

1. 1-丁基-3-甲基咪唑溴化盐（bmimBr）的制备

在配有回流冷凝管、恒压滴液漏斗的 150mL 三口烧瓶中加入 *N*-甲基咪唑 4.1g（50mmol），加热至 70℃，搅拌下缓慢滴加 6.85g（50mmol）溴代正丁烷，滴加结束后继续在 70℃下搅拌反应 3h，得到浅黄色黏稠液体，冷却至室温，用 20mL 乙酸乙酯分三次洗涤，然后减压蒸去溶剂，冷却，得到白色固体，即为 1-丁基-3-甲基咪唑溴化盐。

2. 1-丁基-3-甲基咪唑四氟硼酸盐（$bmimBF_4$）的制备

在 100mL 圆底烧瓶中加入干燥的 1-丁基-3 甲基咪唑溴化盐 5.46g（25mmol），$NaBF_4$ 2.62g（25mmol）和甲醇 20mL，于 40℃搅拌 3h，冷却后抽滤去除 NaBr 沉淀，减压蒸去溶剂，得到浅黄色液体。粗产物中加入 10mL 蒸馏水和 150mL 二氯甲烷，搅拌 10min 后静置分层，收集下层溶液于 250mL 圆底烧瓶中，减压蒸除溶剂得无黏稠液体，即为离子液体 $bmimBF_4$。

五、实验 HSE

1. 甲醇为低毒性、易挥发易燃溶剂。应与强氧化剂、食用化学品等分开存放。保持容器密封，远离火种、热源。在有甲醇蒸气的现场短时间工作时，要佩戴防毒面具、橡皮手套、防护眼镜等安全用具。

2. 二氯甲烷为低毒性、低沸点溶剂，遇明火高热可燃。二氯甲烷直接溅入眼中，有疼痛感并有腐蚀作用，蒸气有麻醉作用。储存于阴凉、干燥、通风良好及阳光无法直射的地方。储存须远离热源、火焰及不相容物，如强氧化剂、强酸。

3. 离子液体制备过程需在通风橱中进行。

六、思考题

1. 合成离子液体的原理是什么？

2. 如何检测 Br^- 是否完全转换为 BF_4^-？

Experiment 19 Synthesis of the ionic liquid bmimBF$_4$

1. Objectives

1.1 Master the synthesis of the ionic liquid bmimBF$_4$.

1.2 Understand the properties and applications of ionic liquids.

2. Principle

Ionic liquids (ILs), known as room temperature molten salt, present a liquid state at or near room temperature (generally $<100℃$). ILs are generally composed of organic cations containing nitrogen and phosphorus and inorganic or organic anions. Normally ionic compounds can only become liquid at high temperatures. However, ionic liquids have a large volume and loose structure, resulting in a low electrostatic force between the cations and the anions. The melting point is close to room temperature.

Compared with traditional organic solvents, ionic liquids have the advantages: wider liquid range, negligible vapor pressure, non-flammable nature, controllable physicochemical properties, wider electrochemical window, recyclability, and so on. Ionic liquids have been widely used as reaction medium and catalyst for organic synthesis, solvent and extractant for separation science, and electrochemical electrolyte. ILs have become an important part of the study of green chemistry.

The ionic liquid 1-butyl-3-methylimidazolium tetrafluoroborate (bmimBF$_4$) is the most common room temperature ionic liquid. Its synthesis route is as follows:

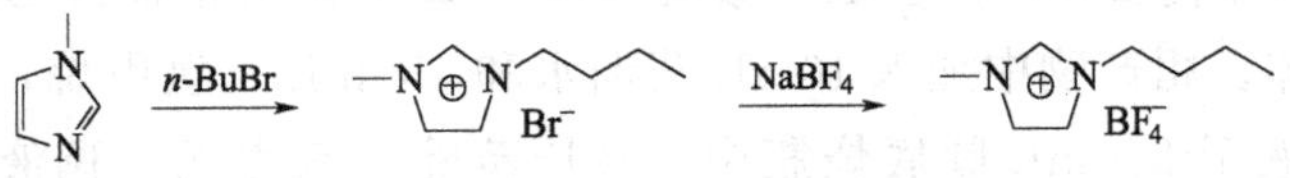

3. Instruments and Reagents

3.1 Instruments

Constant pressure drop funnel, 150mL three-necked flask, 100mL flask, circulating water vacuum pump, thermometers, Büchner funnels, filter bottles, glass rods, separating funnels, electronic balances, heating sets, iron stands, transformers, rotary evaporator.

3.2 Reagents

N-methylimidazole, bromobutane, ethyl acetate, methanol, dichloromethane, Sodium tetrafluoroborate.

4. Procedures

4.1 Preparation of 1- butyl -3- methyl imidazole bromide salt (bmimBr)

Fit a 150mL three-neck flask equipped with a reflux condenser, a constant pressure drop funnel, and a stirring machine. Add 4.1g of *N*-methyl imidazole (50mmol) into the flask, heat to 70℃ and maintain the temperature, begin to drop slowly 6.85g (50mmol) of bromobutane. Then keep on stirring for 3 h at 70℃. When the mixture is cooled to room temperature, wash the mixture with 20mL ethyl acetate three times. Then evaporate the solvent under reduced pressure to get the white solid product.

4.2 Preparation of 1-butyl-3-methyl imidazole tetrafluoroborate ($bmimBF_4$)

Add dry 5.46g (25 mmol) of 1-butyl-3-methylimidazole bromide, 2.62g of $NaBF_4$ (25 mmol), and 20mL methanol in a 100mL round bottom flask, stir at 40℃ for 3h, remove NaBr precipitation by vacuum filtration after cooling, and evaporate solvent under reduced pressure to obtain light yellow liquid. Add 10mL distilled water and 150mL dichloromethane to the crude product, stir for 10 min and then stand to make sure two layers, collect the lower layer into a 250mL round bottom flask, evaporate solvent under reduced pressure to get colorless viscous liquid, namely ionic liquid $bmimBF_4$.

5. Experimental HSE

5.1 Methanol is a low toxic and volatile flammable solvent. It should be stored separately from strong oxidants, edible chemicals, etc. Keep the container sealed and away from kindling and heat sources. Wear gas masks, rubber gloves, protective glasses, and other safety appliances when working for a short time in the field with methanol vapor.

5.2 Dichloromethane is low toxicity and low boiling point solvent, which is highly flammable in case of an open fire. Dichloromethane splashed directly into the eyes could cause pain and corrosion, and steam has an anesthetic effect. Store in a cool, dry, well-ventilated place and out of direct sunlight. Store away from a heat source, flame and incompatibilities, strong oxidant and strong acid.

5.3 The synthesis of ionic liquids needs to be carried out in a fume hood.

6. Questions

6.1 What is the principle of synthesizing of ionic liquids?

6.2 How to detect whether Br^- is fully exchanged into BF_4^- or not?

实验二十 柠檬精油的提取

一、实验目的

1. 了解植物精油的基本知识。
2. 掌握提取植物精油的方法。

二、实验原理

植物精油是从植物的枝、叶、花、果实等部位提取的有机化合物的混合物，主要成分为单萜、芳香烃衍生物等，通常具有挥发性。精油有强烈的特殊香味，极强的渗透力，能够迅速进入人体皮肤，进入血液循环。

柠檬具有许多重要的天然化学物质，包括柠檬酸、抗坏血酸、矿物质、黄酮和精油。柠檬精油是生产食品、日用品、香精、香料的优质原料。

精油的提取和分离的技术主要有水蒸气蒸馏法、溶剂提取法、压榨法、酶提取法、超临界流体萃取法、分子蒸馏法等。

水蒸气蒸馏法是提取精油使用最广泛的方法。组成精油的物质多具有挥发性，可随水蒸气逸出，且冷凝后易与水分离。但提取温度较高，某些成分可能被破坏，因此该法所得的香料的留香性和抗氧化性较差。

压榨法一般用于柑橘类植物精油的提取。利用压榨机将果皮直接冷榨，就可获得粗精油，经分离即可获得精油。此法通常在常温下进行，精油成分很少被破坏，因此可保持天然香味。

三、仪器与试剂

1. 仪器

循环水真空泵，高速离心机，冰箱，研钵，布氏漏斗，离心试管，吸管，250mL 烧瓶，恒压滴液漏斗，冷凝管。

2. 试剂

新鲜柠檬皮，柠檬草。

四、实验步骤

1. 冷榨法提取柠檬油

将新鲜的柠檬皮洗净，晾干 1～2 天备用。

取柠檬皮 50g，切成小颗粒，放入研钵中研磨（或采用小型压榨机）。将榨出物转至布氏漏斗中，抽滤，少量水洗两次后合并滤液，把滤液转移入离心试管中，用高速离心机离心 5min（15000r/min）。将油层用吸管吸出，向残液中加入少量水，再离心一次，合并油相，得到粗油。记录精油的外观及质量。为把粗油中的蜡质分离除去，可把它放入冰箱静置冷藏一周，待杂质下沉后将上层清油吸出，得到较好的冷榨柠檬油。

2. 水蒸气蒸馏法提取柠檬草精油

取柠檬草 100g 洗净，切成 0.5cm 的小段，放入 250mL 的烧瓶中，加水 100mL 和少量沸石。在烧瓶上装上恒压滴液漏斗及回流冷凝管。关闭漏斗下方活塞，加热使瓶内的水保持沸腾。从冷凝管回流下来的液体被收集在恒压滴液漏斗中，分成两相。适时将漏斗下端活塞拧开，将下层的水放回烧瓶中，蒸馏约 2h。降温，分出漏斗内下层的水，将余下部分移入试管中，用高速离心机分离，得到柠檬草精油，记录精油的外观及质量，计算产率。

五、实验 HSE

1. 正确规范使用高速离心机，以免造成人身伤害和仪器损坏。

2. 水蒸气蒸馏法蒸馏过程中需随时观察漏斗中水量，及时打开活塞将下层水放回烧瓶。注意温度较高，小心操作防止烫伤。

六、思考题

1. 比较冷榨法和水蒸气蒸馏法提取的精油色泽、香味有何不同？

2. 查资料，拟出玫瑰精油的提取方法。

Experiment 20 Extraction of lemon essential oil

1. Objectives

1.1 Understand the basics of plant essential oils.

1.2 Master the method of extracting plant essential oils.

2. Principle

Plant essential oil is a mixture of organic compounds extracted from the branches, leaves, flowers, and fruits of plants. The main component of the essential oils is a monoterpene, aromatic hydrocarbon derivatives, etc. The essential oil has a strong special fragrance and strong penetration, can quickly penetrate the human skin and enter into the bloodstream.

Lemons have many important natural chemical components, including citric acid, ascorbic acid, minerals, flavonoids, and essential oils. Lemon oil is an essential oil of considerable commercial importance and is widely used in the flavor and fragrance industry.

The extraction and separation techniques of essential oil mainly include steam distillation, solvent extraction, cold-pressed, enzyme extraction, supercritical fluid extraction (SFE), molecular distillation, and so on. Each of these methods has its own merits and drawbacks.

Hydrodistillation is the most widely used method to extract essential oils. The components of essential oils are volatile, can escape with water vapor, and can be easily separated from the water after condensation. However, due to the high extraction temperature, some components may be destroyed, so the fragrances obtained by this method have poor retention and antioxidant properties.

The cold-pressed is generally used for the extraction of citrus essential oils. The peel of fresh fruits is cold-pressed to obtain the crude essential oil. This method is usually carried out at room temperature, so the essential oil components are rarely destroyed, and the natural fragrance can be maintained.

3. Instruments and Reagents

3.1 Instruments

Circulating water vacuum pump, high-speed centrifuge, refrigerator, mortar, Büchner funnel, centrifugal tube, constant pressure drop funnel, beaker, 250mL flask, straw.

3.2 Reagents

Fresh lemon peel, lemongrass.

4. Procedures

4.1 Extraction of essential oil from lemon peel by a cold-pressed method

Wash fresh lemon peel and let it dry for 1～2 days.

100g lemon peels are cut into small pieces and ground in a mortar (or in a small press). Then the extracted material is transferred to a Büchner funnel to filter and washed with a small amount of water twice, and then combined with the filtrate. The filtrate is transferred into a centrifuge tube and dehydrated at a high-speed centrifuge for 5min (15000r/min). The oil layer is sucked out with a straw, a small amount of water is added to the residual fluid and then centrifuged again. Combine the oil phase to get the crude oil. Record the appearance and quality of crude essential oils. To separate and remove the wax in crude oil, it can be put into the refrigerator for a week to precipitate the impurities. Suck out the upper oil to get better cold-pressed lemon oil.

4. 2 Extraction of lemongrass essential oil by steam distillation

100g of cleaned lemongrass is cut ted into small sections of about 0. 5 cm. Then put them into a 250mL flask with 100mL water and a small amount of zeolite. The flask is fitted with a constant pressure drop funnel and a reflux condenser. Close the piston below the funnel and heat to keep the water in the bottle boiling. The condensed liquid is collected in a constant pressure drop funnel, forming two phases. Unscrew the piston at the bottom of the funnel and return the water to the flask. Distill for about 2 hours. Make it cool and drain water from the bottom of the funnel. The remaining part is transferred into a centrifuge tube and separated by a high-speed centrifuge to obtain lemongrass essential oil. Record the appearance and quality of the essential oil and calculate the yield.

5. Experimental HSE

5. 1 Use high-speed centrifuge correctly to avoid personal injury and instrument damage.

5. 2 In the steam distillation process, it is necessary to observe the water quantity in the funnel at any time, open the piston in time, and put the lower water back into the flask. Pay attention to the high temperature, and operate carefully to prevent scalding.

6. Questions

6. 1 Compare the color and fragrance of essential oil extracted by cold-pressed and steam distillation.

6. 2 Search literature, draw up the extraction method of rose essential oil.

实验二十一 柠檬精油的GC-MS分析

一、实验目的

1. 了解GC-MS仪器的组成、结构和特点。

2. 学习用GC-MS技术对精油成分进行定性、定量分析。

二、实验原理

气相色谱法（GC）是利用不同物质在固定相和流动相分配系数的差别，使不同化合物从色谱柱流出的时间不同，以达到分离目的方法。

质谱法（MS）中，化合物在高真空中受热气化，气态分子受到一定能量的电子流轰击后失去一个价电子成为正电荷的分子离子，分子离子进一步碎裂为碎片离子，利用带电粒子在磁场或电场中的运动规律，按其质荷比实现分离分析，测定离子质量及其强度分布。质谱法能给出化合物的分子量、经验式及分子结构信息，具有定性专属性强、灵敏度高、检测快速的优势。

气相色谱-质谱联用技术（GC-MS）是将气相色谱法与质谱法联用的一种技术，应用十分广泛。质谱法可以进行有效的定性分析，但对复杂有机化合物的分析就显得无能为力；而色谱法对有机化合物是一种有效的分离分析方法，特别适用于进行有机化合物的定量分析，但定性分析则比较困难。因此，这两者的有效结合为化学家及生物化学家提供一个高效的可以进行复杂有机化合物定性、定量的分析工具。

柠檬含有许多重要的天然化学成分，包括柠檬酸、抗坏血酸、矿物质、类黄酮和精油。精油的挥发性部分由碳氢化合物和倍半萜烯组成。柠檬精油的主要成分为柠檬烯，一般含量在60%～70%之间；有较多的γ-松油烯（9%～18%）和α-松萜（7%～11%），含有2.3%～6.3%的氧化物。

三、仪器与试剂

1. 仪器

GC-MS联用仪，0.22μm微孔滤膜。

2. 试剂

柠檬精油，二氯甲烷。

四、实验步骤

1. 样品预处理

将提取的柠檬精油用二氯甲烷稀释10倍，经0.22μm微孔滤膜过滤，待测。

2. 样品分析

气相色谱条件：色谱柱为SE-52毛细管柱（30mm×0.32mm，0.25μm），载气为高纯氦气（99.999%），流速0.8mL/min，分流比20：1，柱温50℃保持5min，再以5℃/min升温至230℃，保持2min。进样温度和检测温度分别为220℃和240℃，进样量0.5μL。

质谱条件：离子源温度230℃，接口温度230℃，EI电子源，电子能量70eV，m/z扫描范围35～500。

五、实验HSE

1. 进样口高温，不得触碰以免烫伤。

2. 二氯甲烷为低毒性、低沸点溶剂，遇明火高热可燃。二氯甲烷直接溅入眼中，有疼痛感并有腐蚀作用，蒸气有麻醉作用。储存于阴凉、干燥、通风良好及阳光无法直射的地方。储存须远离热源、火焰及不相容物，如强氧化剂、强酸、硝酸。

六、思考题

1. 气质联用仪的工作原理是什么？
2. 比较样品图谱和检索库 NIST 库中图谱。

Experiment 21 GC-MS analysis of lemon essential oil

1. Objectives

1.1 Understand the structure, composition, and characteristics of the GC-MS instrument.

1.2 Learn to use GC-MS for qualitative and quantitative analysis of constituents of essential oil.

2. Principle

Gas chromatography (GC) is a separation method that uses the difference in the distribution coefficients of different substances between the stationary phase and mobile phase to make different compounds flow out of the column at different times.

Mass spectrometry (MS) is a qualitative analysis method. The compound is vaporized in a high vacuum. The gaseous molecule is bombarded by certain energy of electron flow and loses a valence electron to become a positively charged molecular ion. The molecular ion is further broken into fragments. The mass and intensity distribution of ions is determined by the separation and analysis of charged particles in a magnetic field or electric field according to their mass to charge ratio. Mass spectrometry can give the molecular weight, empirical formula, and molecular structure information of compounds, and has the advantages of strong qualitative specificity, high sensitivity, and fast detection.

Gas chromatography-mass spectrometry (GC-MS) is a technique that combines gas chromatography-mass spectrometry (GC-MS) with mass spectrometry. Mass spectrometry can perform effective qualitative analysis, but it is powerless to analyze complex organic compounds; while chromatography is an effective separation analysis method for organic compounds, especially suitable for quantitative analysis of organic compounds, but qualitative analysis is more difficult. Therefore, the effective combination of the two method provides chemists and biochemists with an efficient qualitative and quantitative analysis tool for complex organic compounds.

Lemons have many important natural chemical components, including citric acid, ascorbic acid, minerals, flavonoids, and essential oils. The volatile fraction of essential oils are made up of hydrocarbons and sesquiterpene. The main component of essential oils is limonene, generally present between 60% and 70%; there are appreciable amounts of γ-terpinene (9%~18%) and α-terpinene (7%~11%) while oxygenated compounds make up between 2.3% and 6.3%.

3. Instruments and Reagents

3.1 Instruments

GC-MS instrument，0.22μm millipore filter.

3.2 Reagents

Lemon essential oil，dichloromethane.

4. Procedures

4.1 Sample pretreatment

The extracted lemon essential oil was diluted 10 times with dichloromethane and filtered through a 0.22μm microporous filter membrane to be tested.

4.2 Sample analysis

Gas chromatographic conditions：the chromatographic column was SE-52 capillary column (30mm×0.32mm，0.25μm). The carrier gas was high purity helium (99.999%). The flow rate was 0.8mL/min，the split ratio was 20∶1. The oven temperature was programmed from 50℃ (5 min isothermal) to 230℃ at 5℃/min. The temperatures of the injector and detector were 220℃ and 240℃，respectively. The injection volume was 0.5μL.

Mass spectrum conditions：ion source temperature 230℃，interface temperature 230℃，EI electron source，electron energy 70eV，*m/z* scanning range of 35～500.

5. Experimental HSE

5.1 High temperature of sample inlet，do not touch to avoid scalding.

5.2 Dichloromethane is low toxicity solvent with low boiling point，which is highly flammable in case of an open fire. Dichloromethane splashed directly into the eyes could cause pain and corrosion，and steam has an anesthetic effect. Store in a cool，dry，well-ventilated place and out of direct sunlight. Store away from a heat source，flame and incompatibilities，strong oxidant，strong acid，and nitric acid.

6. Questions

6.1 What is the working principle of GC-MS?

6.2 Analyse the spectra of the sample by comparing the standard ones in the NIST retrieval library.

实验二十二 钛硅分子筛的合成

一、实验目的

1. 了解钛硅分子筛的性质及应用。
2. 掌握水热合成法制备钛硅分子筛的技术。
3. 理解钛硅分子筛的催化方式。

二、实验原理

钛硅分子筛（TS-1）是一种具有 MFI 结构的分子筛，由于其优异的催化氧化性能被广泛地应用于石油化工以及精细化工等领域，TS-1分子筛由于其分子筛骨架当中的T-O四面体，在一些液相催化氧化反应体系中展现了良好的催化活性，尤其是TS-1/双氧水体系，广泛地应用于苯（苯酚）的羟基化反应、酮的氨肟化反应、烯烃的环氧化反应等诸多利用双氧水作为氧化剂来催化的氧化反应中，且副产物仅有水。

自从M. Taramasso等将钛原子引入纯硅分子筛Silicalite-1当中，首次合成Titanium Silicalite-1（TS-1）以来，TS-1分子筛凭借其优良的催化性能受到了广泛研究。TS-1分子筛具有二维十元环孔道，其中一个为平行于b轴方向的直孔道，另一个为Zigzag型十元环孔道，水热合成法和同晶取代法是合成TS-1分子筛最常用的两种方法，其中对合成分子筛最显著的影响有：硅源、钛源、模板剂以及晶化条件。

水热合成法是制备TS-1的方法中极其重要的一种方法，它是以模板剂、硅源、钛源在高温水热条件下进行水热晶化合成的纯度较高的结晶粉末。水热合成法的主要特点有：在水溶液中离子混合均匀；水随温度升高和压力增大变成一种气态矿化剂，具有非常大的解聚能力。水热物系在一定矿化剂存在下，化学反应速度快，能制备出多组分或单一组分的超微结晶粉末；离子能够比较容易地按照化学计量反应，晶粒按其结晶习性生长，在结晶过程中，可把有害杂质自排到溶液当中，生成纯度较高的结晶粉末。

水热合成法的优点是所得产物纯度高、分散性好、粒度易控制。该工艺原料的纯度较高，整个过程不引入杂质离子，可以通过严格控制工艺条件，制得纯度高、粒径小、粒度分布窄的纳米粉体，且产品质量稳定。

本实验以正硅酸四乙酯为硅源，四丙基氢氧化铵为模板剂，钛酸四丁酯为钛源，通过水热合成法制备钛硅分子筛（TS-1）。

三、仪器与试剂

1. 仪器

磁力搅拌器，烧杯，马弗炉，离心机，高压反应釜。

2. 试剂

四丙基氢氧化铵，去离子水，正硅酸四乙酯，钛酸四丁酯，异丙醇。

四、实验步骤

取6.4g四丙基氢氧化铵（TPAOH）于烧杯中，剧烈搅拌下滴加11.5g正硅酸四乙酯（TEOS）。将0.55g钛酸四丁酯溶于2.5g异丙醇中，并于剧烈搅拌下滴加到反应烧杯中。当溶液搅拌澄清时，再滴加2.4g TPAOH。加入17mL蒸馏水，将混合物加热至353K，并保持3h，以除去体系中的醇。在加热过程中，补加适量的蒸馏水，以保持总体积不变。除醇结束后，将混合物转移到带四氟乙烯衬里的水热反应釜中，并在443K下晶化48h。过滤，用蒸馏水洗涤，干燥后，在823K下焙烧6h，得到TS-1。

五、实验HSE

1. 蒸除异丙醇要小心操作，以防异丙醇蒸气在空气中燃烧。

2. 焙烧使用高温马弗炉，按规定操作，小心烫伤。

3. 四丙基氢氧化铵：密度：1.00g/mL（20℃），沸点：100～102℃，闪点：102℃，折射率：n_{20}^{D} 1.370，储存条件：2～8℃。急性毒性：造成严重皮肤灼伤和眼损伤。造成严重眼损伤。急救措施：吸入，请将患者移到新鲜空气处；皮肤接触，脱去污染的衣着，用肥皂水和清水彻底冲洗皮肤，如有不适感，就医；眼睛接触，分开眼睑，用流动清水或生理盐水冲洗，立即就医；食入，漱口，禁止催吐，立即就医。储存注意事项：储存于阴凉、通风的库房。库温不宜超过 37℃。应与氧化剂、食用化学品分开存放，切忌混储。保持容器密封。远离火种、热源。库房必须安装避雷设备。排风系统应设有导除静电的接地装置。采用防爆型照明、通风设置。禁止使用易产生火花的设备和工具。储区应备有泄漏应急处理设备和合适的收容材料。

4. 正硅酸四乙酯：乙酸硅（Ⅳ）又称四乙酰氧基硅烷，分子式是 $C_8H_{12}O_8Si$，分子量为 264.2616，该物质会引起灼伤。急性毒性：造成严重皮肤灼伤和眼损伤。急救措施：吸入，请将患者移到新鲜空气处；皮肤接触，脱去污染的衣着，用肥皂水和清水彻底冲洗皮肤，如有不适感，就医；眼睛接触，分开眼睑，用流动清水或生理盐水冲洗，立即就医；食入，漱口，禁止催吐，立即就医。储存注意事项：储存于阴凉、通风的库房。库温不宜超过 37℃。应与氧化剂、食用化学品分开存放，切忌混储。保持容器密封。远离火种、热源。库房必须安装避雷设备。排风系统应设有导除静电的接地装置。采用防爆型照明、通风设置。禁止使用易产生火花的设备和工具。储区应备有泄漏应急处理设备和合适的收容材料。

5. 钛酸四丁酯：相对密度：0.996，凝固点：－55℃，沸点：310～314℃，折射率：1.486～1.492，闪点：76.7℃，外观：无色至淡黄色透明液体，溶解性：能溶于除酮以外的大部分有机溶剂，稳定性：置于空气中易固化为透明细片，遇水分解，易燃，毒性：低毒。急性毒性：易燃液体和蒸气。造成皮肤刺激。造成严重眼损伤。可引起呼吸道刺激。可引起昏睡或眩晕。急救措施：吸入，请将患者移到新鲜空气处；皮肤接触，脱去污染的衣着，用肥皂水和清水彻底冲洗皮肤，如有不适感，就医；眼睛接触，分开眼睑，用流动清水或生理盐水冲洗，立即就医；食入，漱口，禁止催吐，立即就医。储存注意事项：储存于阴凉、通风的库房。库温不宜超过 37℃。应与氧化剂、食用化学品分开存放，切忌混储。保持容器密封。远离火种、热源。库房必须安装避雷设备。排风系统应设有导除静电的接地装置。采用防爆型照明、通风设置。禁止使用易产生火花的设备和工具。储区应备有泄漏应急处理设备和合适的收容材料。

6. 异丙醇：无色透明液体，有似乙醇和丙酮混合物的气味。溶于水，也溶于醇、醚、苯、氯仿等多数有机溶剂。异丙醇是重要的化工产品和原料，主要用于制药、化妆品、塑料、香料、涂料等。急性毒性：高度易燃液体和蒸气。造成严重眼刺激。可引起昏睡或眩晕。急救措施：吸入，迅速脱离现场至空气新鲜处，保持呼吸道通畅，如呼吸困难给输氧，如呼吸、心跳停止立即进行心肺复苏术，就医；皮肤接触，立即脱去污染的衣着，用肥皂水和清水彻底冲洗，就医；眼睛接触，立即分开眼睑，用流动清水或生理盐水彻底冲洗，就医，食入，漱口，不要催吐，就医，经口摄入大量异丙醇者应在 30min 内洗胃，因异丙醇快速被吸收，2h 后洗胃无益，可做血液透析。酒精增加本品毒性。储存注意事项：耐火设备（条件）。与强氧化剂分开存放。阴凉场所。严格密封。

六、思考题

1. 四丙基氢氧化铵在反应中发挥怎样的作用？

2. 加热过程中为什么要补充蒸馏水来保证体积？

Experiment 22 Synthesis of titanium silicalite molecular sieve(TS-1)

1. Objectives

1.1 Understand the properties and application of titanium silicalite molecular sieve.

1.2 Master the technology of hydrothermal synthesis of titanium silicalite molecular sieve.

1.3 Learn to understand the catalytic way of titanium silicalite.

2. Principle

TS-1 is a kind of molecular sieve with MFI structure. Because of its excellent catalytic oxidation performance, it has been widely used in the petrochemical industry and fine chemical industry. Because of the T-O tetrahedron in its molecular sieve framework, TS-1 molecular sieve has shown good catalytic activity in some liquid-phase catalytic oxidation reaction systems, especially the TS-1/hydrogen peroxide system has been widely used in oxidation reactions which use hydrogen peroxide as oxidants, such as hydroxylation reaction of benzene (phenol), ammoximation reaction of ketone, epoxidation reaction of olefin, and the by-product is only water.

Since M. Taramasso et al. titanium silicalite-1 (TS-1) has been synthesized for the first time by introducing titanium atom into silicalite-1, TS-1 has been widely studied for its excellent catalytic performance. TS-1 molecular sieve has two-dimensional ten membered ring channels, one of which is parallel to the b-axis, and the other is a zigzag ten membered ring channel. Hydrothermal synthesis and isomorphic substitution are the two most commonly methods for the synthesis of TS-1 molecular sieve, among which the most significant effects on the synthesis of the molecular sieve are silicon source, titanium source, template agent, and crystallization conditions.

Hydrothermal synthesis is one of the most important methods in the preparation of TS-1. It uses a template, silicon source and titanium source to hydrothermal synthesize crystalline powder with high purity under high-temperature hydrothermal conditions. The main characteristics of hydrothermal synthesis are: the ion is mixed uniformly in the aqueous solution; water becomes a kind of gaseous mineralizer with the increase of temperature and autogenous pressure, which has a very large depolymerization ability. In the presence of a certain mineralizer, the hydrothermal system has a fast chemical reaction speed and can prepare multi-component or single component ultra-fine product powder; the ions can react easily according to the stoichiometry, and the grains grow according to their crystallization habits. During the crystallization process, harmful impurities can be discharged into the solution automatically to generate crystal powder with high purity.

The advantages of hydrothermal synthesis are high purity, good dispersion, and easy controled of particle size. The purity of raw materials is high, and no impurity ions are introduced into the whole process. By strictly controlling the process conditions, nanometer powders with high purity, small particle size, and narrow particle size distribution can be prepared, and the product quality is stable.

In this experiment, tetraethyl orthosilicate was used as a silicon source, tetrapropylammonium hydroxide as a template, and tetrabutyl titanate as a titanium source. Titanium-silicon molecular sieve (TS-1) was prepared by hydrothermal synthesis.

3. Instruments and reagents

3.1 Instruments

Magnetic stirrer, beaker, muffle furnace, centrifuge, high-pressure reactor.

3.2 Reagents

Tetrapropylammonium hydroxide, deionized water, tetraethyl orthosilicate, tetrabutyl titanate, isopropanol.

4. Procedures

Take 6.4g of tetrapropylammonium hydroxide (tpaoh) in the beaker and add 11.5g of tetraethyl orthosilicate (TEOS) under intense stirring. Dissolve 0.55g of tetrabutyl titanate in 2.5g of isopropanol, and drop it into the reaction beaker under intense stirring. When the solution is stirred and clarified, add 2.4g tpaoh. Add 17mL of distilled water, heat the mixture to 353K and hold for 3h to remove the alcohol from the system. In the heating process, add some distilled water to keep the total volume unchanged. After dealcoholization, the mixture was transferred to a hydrothermal reactor lined with tetrafluoroethylene and crystallized at 443K for 48h. After filtration, washing with distilled water, drying, roasting at 823K for 6h, TS-1 was obtained.

5. Experimental HSE

5.1 Tetrapropyl ammonium hydroxide: density: 1.00g/mL at 20℃, boiling point: 100～102℃, flash point: 102℃, refractive index: n_{20}^{D} 1.370, storage conditions: 2～8℃. Acute toxicity: causes severe skin burns and eye damage. Cause severe eye damage. First aid: Inhalation: If inhaled, move the victim to fresh air. Skin contact: remove contaminated clothing and rinse thoroughly with soap and water. If you feel uncomfortable, seek medical advice. Eye contact: separate eyelids and rinse with running water or normal saline. Seek medical attention immediately. Ingestion: gargle, do not induce vomiting. Seek medical attention immediately. Storage precautions: Store in a cool and ventilated warehouse. The temperature should not exceed 37℃. Should be stored separately with an oxidant, edible chemical, avoid mixed storage by all means. Keep container airtight. Keep away from fire and heat. The warehouse must be equipped with lightning protection equipment. The exhaust air system shall be equipped with a ground device to conduct electrostatic discharge. Explosion-proof lighting and ventilation are adopted. Do not use spark generating equipment and tools. The storage area should be equipped with emergency spill handling equipment and suitable storage materials.

5.2 Tetraethyl orthosilicate: silicon acetate (Ⅳ), also known as tetraacetyloxysilane, has a molecular formula of $C_8H_{12}O_8Si$ and a molecular weight of 264. 2616, which can cause burns. Acute toxicity: causes severe skin burns and eye damage. First aid: inhalation: if inhaled, move the victim to fresh air. Skin contact: remove contaminated clothing and rinse thoroughly with soap and water. If you feel uncomfortable, seek medical advice. Eye contact: separate eyelids and rinse with running water or normal saline. Seek medical attention immediately. Ingestion: gargle, do not induce vomiting. Seek medical attention immediately. Storage precautions: Store in a cool and ventilated warehouse. The temperature should not exceed 37℃. Should be stored separately with an oxidant, edible chemical, avoid mixed storage by all means. Keep container airtight. Keep away from fire and heat. The warehouse must be equipped with lightning protection equipment. The exhaust air system shall be equipped with a ground device to conduct electrostatic discharge. Explosion-proof lighting and ventilation are adopted. Do not use spark generating equipment and tools. The storage area should be equipped with emergency spill handling equipment and suitable storage materials.

5. 3 Tetrabutyl titanate: relative density: 0. 996, freezing point, and 55℃, boiling point: 310 ～ 314℃, the refractive index: 1. 486～1. 492, the flashpoint: 76. 7℃, appearance: colorless to light yellow transparent liquid, solubility: soluble in most organic solvents, except ketone stability: easy cure for transparent fritter placed in the air, and water decomposition, flammable, toxic, low toxicity. Acute toxicity: flammable liquids and vapors. Cause skin irritation. Cause severe eye damage. Can cause respiratory irritation. Can cause lethargy or dizziness. First aid: inhalation: if inhaled, move the victim to fresh air. Skin contact: remove contaminated clothing and rinse thoroughly with soap and water. If you feel uncomfortable, seek medical advice. Eye contact: separate eyelids and rinse with running water or normal saline. Seek medical attention immediately. Eat into the mouth rinse, prohibit to induce vomiting. Seek medical attention immediately. Storage precautions: Store in a cool and ventilated warehouse. The temperature should not exceed 37℃. Should be stored separately with an oxidant, edible chemical, avoid mixed storage by all means. Keep container airtight. Keep away from fire and heat. The warehouse must be equipped with lightning protection equipment. The exhaust air system shall be equipped with a ground device to conduct electrostatic discharge. Explosion-proof lighting and ventilation are adopted. Do not use spark generating equipment and tools. The storage area should be equipped with emergency spill handling equipment and suitable storage materials.

5. 4 Isopropanol: a clear, colorless liquid with a smell like a mixture of ethanol and acetone. Soluble in water, but also soluble in alcohol, ether, benzene, chloroform, and other organic solvents. Isopropanol is an important chemical product and raw material. Mainly used in pharmaceutical, cosmetic, plastic, spice, paint, etc. Acute toxicity: highly flammable liquids and vapors. Cause severe eye irritation. Can cause lethargy or dizziness. First-aid measures: inhalation: quickly leave the scene to the fresh air. Keep your airway open. If breathing is difficult, give oxygen. Breathing, cardiac arrest, cardiopulmonary resuscitation immediately. Go to a doctor. Skin contact: immediately remove contaminated clothing and

rinse thoroughly with soap and water. Go to a doctor. Eye contact: immediately separate eyelids and rinse thoroughly with running water or normal saline. Go to a doctor. Ingestion: gargle and do not induce vomiting. Go to a doctor. People who take a large amount of isopropanol orally should have their stomach washed within 30 minutes. You can do hemodialysis. Alcohol increases the toxicity of this product. Storage precautions: fire-resistant equipment (condition). Store separately from a strong oxidizer. A cool place. Tightly sealed.

6. Questions

6.1 What role does tetrapropylammonium hydroxide played in the reaction?

6.2 Why is it necessary to add distilled water to maintain volume during heating?

实验二十三 钛硅分子筛的表征(SEM)

一、实验目的

1. 掌握 SEM 样品的常规制备技术。
2. 熟悉 SEM 测试的基本操作方法。
3. 了解 TS-1 的晶体形貌。

二、实验原理

扫描电子显微镜（Scanning electron microscope，SEM），是一种有效的显微结构分析工具，主要利用二次电子信号成像来观察样品的表面形态，即用极狭窄的电子束去扫描样品，通过电子束与样品的相互作用产生各种效应，其中主要是样品的二次电子发射。经检测、放大和信号处理，把数据特征直观地在显示屏上表现出来。它具有制样简单、分辨率超高、景深大、成像立体感强和连续可调等优点，在材料科学研究领域中表面形貌、断口分析等方面显示越来越强的优越性。

扫描电镜用于成像的信号来自入射光束与样品中不同深度的原子的相互作用。样品在电子束的轰击下会产生包括背散射电子、二次电子、特征 X 射线、吸收电子、透射电子、俄歇电子、阴极荧光、电子束感生效应等在内的多种信号，而一个单一机器能够配有所有信号的探测器是很难的，背散射电子（BSE）、二次电子（SEI）、特征 X 射线探测器是一般扫描电子显微镜的标配探测器。

二次电子形貌衬度：二次电子是电子束轰击样品，使样品中原子的外层电子与原子脱离，产生的一种自由电子。二次电子的能量较低，一般在 50eV 以下。由于二次电子产生于距离样品表面很近的位置（一般距表层 5～10nm），二次电子成像（SEI）可以对样品表面进行高分辨率的表征，分辨率可以达到 1nm。因此二次电子模式可用于观察表面形貌特征。

背散射电子原子序数衬度：背散射电子（BSE）是电子束轰击样品过程中被样品反射回来的部分电子，其中包括被原子核反射回来的弹性背散射电子和被原子核外电子反射回来的非弹性背散射电子。弹性背散射电子的散射角大于 90°，没有能量损失，因此弹性背散射电子的能量很高，一般可以达到数千伏到数万伏。非弹性背散射电子由于和核外电子碰撞，不仅方向改变，也会有不同程度的能量损失，因此非弹性背散射电子的能量分布范围较广，一般为数十电子伏到数千电子伏。由于非弹性背散射电子需要经过多次散射才能逸出样品表面，弹性背散射电子的数量是远高于非弹性背散射电子的，因此扫描电镜中所指的背散射电子多指弹性背散射电子。背散射电子产生于距离样品表面几百纳米的深度，因此背散射电子图像的分辨率低于二次电子图像分辨率。然而，背散射电子的产量与样品原子序数有很大的关系，因此可以提供样品原子序数衬度信息。在背散射模式下，样品表面平均原子序数的分辨率低于二次电子图像分辨率。然而，背散射电子的产量与样品原子序数有很大的关系，因此可以用来提供样品原子序数衬度信息。在背散射模式下，样品表面平均原子序数大的区域，背散射信号强，则电镜图中表现为亮度高，相反，原子序数小的区域比较暗。

特征 X 射线：当高能电子束轰击样品，将样品中原子的内层电子电离，此时的原子处于较高激发态，外层的高能量电子会向内层跃迁以填补内层空缺，从而释放能量，这部分辐射能量称为特征 X 射线。这些特征 X 射线可以用来鉴别组成以及测定样品中元素的含量。

三、仪器与试剂

1. 仪器

扫描电子显微镜（SEM），洗耳球，导电胶。

2. 试剂

钛硅分子筛（TS-1）。

四、实验步骤

1. 制样准备

(1) 将导电胶粘贴在样品柱上。

(2) 将隔膜样品切成小片，平贴在导电胶上。

(3) 将少量粉末撒在导电胶上，用洗耳球轻轻吹走没有粘住的样品。

(4) 喷金。

① 将样品柱放入喷金仪中（切记：样品座不可放入）。

② 打开电源（已默认为自动模式），当显示屏数值降低到“6”时，按下“START”开始自动喷金。

③ 喷金完成后，取出样品柱，并将其固定在样品座上，完成制样。

2. 拍摄过程

(1) 开机，将钥匙旋到“START”。

(2) 启动电脑及“InTouchScope”软件。

(3) 样品放置。

① 点击软件上“VENT”，进行放气，当“VENT”常亮时，表示放气完毕。

② 打开样品仓，将样品放在样品台上（暂不关闭样品仓）。

③ 点击软件界面上“Change”图标，弹出对话框（两个选项：一是“Set spectimen”选项，二是观察设置选项），点击“Set spectimen”选项。

④ 弹出对话框，点击“Acquire image”，获取图像。

⑤ 合上样品仓，点击“Provide condition”，进行抽气，抽气完成后进入观察阶段。

(4) 样品观察。

① 选择观察的样品及观察点。

② 交叉调节“MAGNIFICATIAO（倍率）”“FOCUS（焦距）”和“STIGMA（像散）”直到获得满意的图像效果，点击“PHOTO（拍照）”获取图片。

③ 点击软件工具栏上的“File”“Save image file”，进行保存图片。（注意：隔膜观察束斑值“SS”为 30，加速电压“Acc voltage”为 20kV，工作距离“WD”为 10mm）。

（5）关闭仪器

① 拍摄完成后关闭灯丝，并点击软件上“VENT”，进行放气，完成后打开样品仓，取出样品。

② 关闭样品仓，并点击软件上“EVAC”，进行抽真空，使其保持真空状态。

③ 关闭电脑软件及仪器，断开电源。

五、实验 HSE

1. 使用电镜时，首先要观察电镜的状态。

2. 样品的状态必须是干燥的，不含有水分或挥发性物质；具有一定的机械强度，能耐受电子束轰击；具有导电性，被激发时能够产生足够的二次电子；样品大小与厚度要适于样品台的安装；观察面清洁，无污染物。

3. 非导体样品要进行导电处理，即喷镀一层金属膜，一般膜厚为 10～20nm。并根据样品表面凹凸粗糙状况及样品性质而变化。

4. 粉体可以直接撒在试样座的双面碳导电胶上，例如玻璃板压紧，然后用洗耳球吹去黏结不牢固的颗粒；块状试样可直接研磨和抛光，或镶嵌后研磨、抛光。

六、思考题

1. SEM 的工作原理是什么？

2. SEM 的使用范围和局限性是什么？

Experiment 23 Characterization of titanium silicalite molecular sieve TS-1 (SEM)

1. Objectives

1.1 Master the conventional preparation technology of SEM samples.

1.2 Familiar with the basic operation method of SEM test.

1.3 Understand the crystal morphology of TS-1.

2. Principle

A scanning electron microscope (SEM) is an effective tool for microstructure analysis. It mainly uses secondary electron signal imaging to observe the surface morphology of the sample. That is, scanning the sample with an extremely narrow electron beam, which produces various effects through the interaction between the electron beam and the sample, mainly the secondary electron emission of the sample. After detection, video amplification, and signal processing, the data characteristics are directly displayed on the display screen. It has the advantages of simple sample preparation, super-high resolution, large depth of field, strong stereoscopic imaging, and continuously adjustable. It has more and more advantages in the field of material science research, surface morphology, fracture analysis, and so on.

The signal of scanning electron microscopy for imaging comes from the interaction between the incident beam and atoms of different depths in the sample. Under the bombardment of the electron beam, the sample will produce a variety of signals, including backscattered electrons, secondary electrons, characteristic X-rays, absorption electrons, transmission electrons, Auger electrons, cathodoluminescence, electron beam induced effects, etc., but it is difficult for a single machine to be equipped with a detector with all the signals, backscattered electron (BSE), secondary electron (SEI) and characteristic X-ray detector are standard detectors for general scanning electron microscopy.

Principle of secondary electron morphology contrast: the secondary electron is a kind of free electron produced by bombarding the sample with an electron beam to separate the outer electrons of the atoms in the sample from the atoms. The energy of the secondary electron is lower, generally below 50eV. Because the secondary electrons are generated near the surface of the sample (generally 5～10nm away from the surface), the secondary electron imaging (SEI) can characterize the sample surface with a high resolution of 1nm. Therefore, the secondary electron mode is used to observe the surface morphology characteristics.

Principle of atomic number contrast of backscattered electrons: backscattered electrons (BSE) are some of the electrons reflected by the sample in the process of electron beam bombardment, including the elastically backscattered electrons reflected by the atomic nucleus and the inelastic backscattered electrons reflected by the electrons outside the atomic nucleus. The scattering angle of elastically backscattered electrons is greater than 90°, and there is no energy loss. Therefore, the energy of elastically backscattered electrons is very high, which can reach thousands to tens of thousands of volts. Because of colliding with the electrons outside the nucleus, the inelastic backscattered electrons will not only change the direction but also have different degrees of energy loss. Therefore, the energy distribution of inelastic backscattered electrons has a wide range, generally tens to thousands of electron volts. Because inelastic backscattered electrons need to be scattered many times to escape from the surface of the sample, the number of elastically backscattered electrons is much higher than that of inelastic backscattered electrons, so the backscattered electrons in SEM refer to elastically backscattered electrons. The backscattered electrons are generated at a depth of several hundred nanometers from the sample surface, so the resolution of the backscattered electron image is lower than that of the secondary electron image. However, the production of backscattered electrons is closely related to the atomic number of the sample, so it can be used to provide the contrast information of the atomic number of the sample. In the backscattering mode, the resolution of the average atomic number of the sample surface is lower than that of the secondary electron image. However, the production of backscattered electrons is closely related to the atomic number of the sample, so it can be used to provide the contrast information of the atomic number of the sample. In the backscatter mode, the region with a large average atomic number and strong backscatter signal on the sample surface is characterized by high brightness in the electron micrograph, on the contrary, the region with a small atomic number is relatively dark.

Characteristic X-ray principle: when the high-energy electron beam bombards the sample and ionizes the inner electrons of the atoms in the sample, the atoms are in a higher excited state at this time, and the high-energy electrons in the outer layer will leap to the inner layer to fill the inner layer vacancy and release energy, which is called characteristic X-ray. These characteristic X-rays can be used to identify the components and determine the rich elements in the samples.

3. Instruments and Reagents

3.1 Instruments

Scanning electron microscope (SEM), pipette bulb, conductive adhesive.

3.2 Reagents

Titanium silicalite (TS-1).

4. Procedures

4.1 Sample preparation

(1) Paste the conductive adhesive on the sample column.

(2) Cut the diaphragm sample into small pieces and stick it flat on the conductive adhesive; (diaphragm).

(3) Sprinkle a small amount of powder on the conductive adhesive, and gently blow away the sample without sticking with the pipette bulb.

(4) Spraying gold.

① Put the sample column into the gold sprayer (remember: the sample holder cannot be put in).

② Turn on the power (the default mode is auto mode), when the display value is reduced to "6", press "start" to start auto gold spraying.

③ After the gold spraying, take out the sample column and fix it on the sample seat to complete the sample preparation.

4.2 Shooting process

(1) Turn on the machine and turn the key to "START".

(2) Start the computer and "InTouchScope" software.

(3) Sample placement.

① Click "VENT" on the software to deflate. When "VENT" is always on, it means the deflate is completed.

② Open the sample bin and place the sample on the sample table (do not close the sample bin temporarily).

③ Click the "Change" icon on the software interface to pop up the dialog box (two options: one is the "Set spectrum" option, the other is the observation setting option), and click the "Set spectrum" option.

④ Pop up the dialog box, click "Acquire image" to obtain the image.

⑤ Close the sample chamber, click "Provide condition" to pump air, and enter the observation stage after the completion of pumping.

(4) Sample observation.

① Select the observation samples and observation points.

② Cross adjust "MAGNIFICATION", "FOCUS" and "STIGMA" until satisfactory image effect is achieved, and click "PHOTO" to obtain image.

③ Click "File" and "Save image file" on the software toolbar to save the picture. (Note: the diaphragm observation beam spot value "SS" is 30, the acceleration voltage "Acc voltage" is 20kV, and the working distance "WD" is 10mm).

(5) Turn off the instrument.

① After shooting, turn off the filament, click "VENT" on the software, deflate, open the sample bin and take out the sample.

② Close the sample bin and click "EVAC" on the software to vacuum and keep it in a vacuum state.

③ Turn off the computer software and instruments and disconnect the power supply.

5. Experimental HSE

5.1 When using an electron microscope, the state of the electron microscope should be observed first.

5.2 The sample must be dry and free of moisture or volatile substances; it has certain mechanical strength and can withstand electron beam bombardment; it has conductivity and can generate enough secondary electrons when excited; the sample size and thickness shall be suitable for the installation of the sample table; the observation surface shall be clean and free of pollutants.

5.3 The non-conductive samples should be treated with conductive treatment, i. e. spraying a layer of a metal film, generally, the thickness of the film is 10 ~ 20nm. It also needs to be changed according to the roughness of the surface and the nature of the sample.

5.4 The powder can be directly spread on the double-sided carbon conductive adhesive of the sample holder, such as glass plate pressing, and then blow off the loose particles with the pipette bulb; the block sample can be directly ground and polished, or ground and polished after inlaying.

6. Questions

6.1 What is the working principle of SEM?

6.2 What is the using scope and limitations of SEM?

实验二十四　钛硅分子筛催化苯酚制备苯二酚

一、实验目的

1. 了解 TS-1 催化剂的原理和应用。

2. 理解 TS-1 催化剂催化苯酚制备苯二酚反应的原理。

3. 掌握气质联用在反应监控和产物副产物定性中的作用。

二、实验原理

苯酚主要用于生产酚醛树脂和己内酰胺等重要化学品，苯二酚包括邻苯二酚和对苯二酚，广泛应用在香料、染料和橡胶合成等领域。至今为止，世界上 92% 的苯酚由异丙苯法生产，然而异丙苯法对设备腐蚀严重，并且污染环境。苯二酚的生产工艺主要采用苯酚双氧水羟基化法，反应以无机酸盐为催化剂，存在催化剂分离困难、转化率低等诸多问题。研究温和条件下苯酚直接羟基化制苯二酚对开发苯酚/苯二酚的绿色生产工艺具有重要的理论和应用价值。

在水溶剂中，过氧化氢与钛硅分子筛中的四配位骨架钛活性中心相互作用，生成 Ti-OOH 活性物种。随后，游离在溶剂中的苯酚进攻 Ti-OOH，形成苯二酚的中间体，接着苯环上的氢原子与 Ti-OOH 中的羟基交换，生成苯二酚的同时恢复了催化剂的活性，根据邻对位定位规律，主要生成邻苯二酚和对苯二酚，以及少量间苯二酚，随着反应时间的延长，对苯二酚可能会进一步被氧化成苯醌（溶液颜色变深），其反应方程式如图 1 所示。

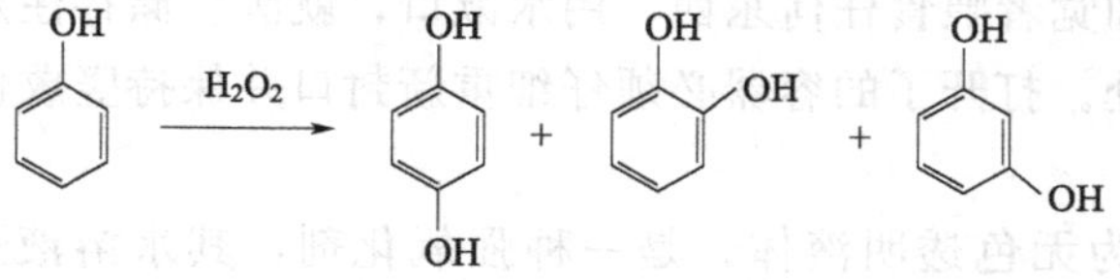

图 1 反应方程式

三、仪器与试剂

1. 仪器

三口圆底烧瓶，磁子，电热套，回流冷凝管，温度计，滴液漏斗，梨形漏斗，气相色谱仪（GC）。

2. 试剂

苯酚，TS-1，双氧水（30%），甲醇，乙酸乙酯。

四、实验步骤

1. 实验装置

2. 实验步骤

如图 2 顺序搭好反应装置，在 100mL 的三口烧瓶中，加入 10g 苯酚，1g TS-1 分子筛，30mL 乙腈，80℃下反应，每隔 0.5h 滴加 2mL 30% 的双氧水，共计加入 8mL 30% 的双氧水，加完双氧水在 80℃继续反应 2h。反应结束后先加少量甲醇淬灭反应，过滤，滤液经气相色谱-质谱联用仪（GC-MS）进行定性、定量分析。

五、实验 HSE

1. 苯酚：苯酚，又名石碳酸、羟基苯，是最简单的酚类有机物，一种弱酸。常温下为一种无色晶体，有毒。苯酚是一种常见的化学品，是生产某些树脂、杀菌剂、防腐剂以及药物（如阿司匹林）的重要原料。苯酚有腐蚀性，常温下微溶于水，易溶于有机溶液；当温度

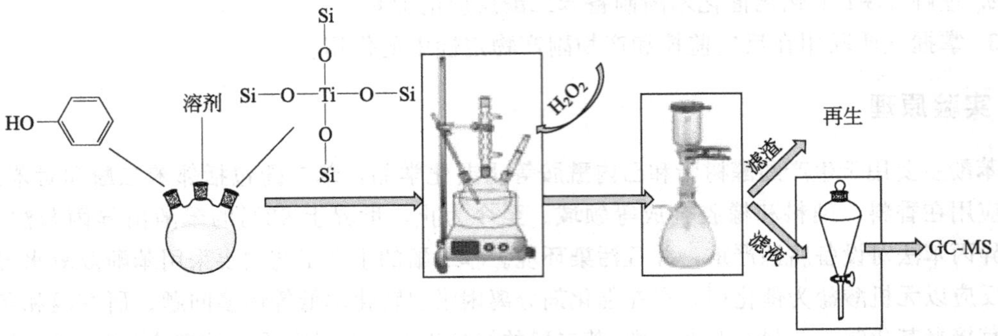

图 2 实验装置

高于 65℃时，能跟水以任意比例互溶。其溶液沾到皮肤上可用酒精洗涤，苯酚暴露在空气中呈粉红色。急性毒性：吞咽会中毒，吸入会中毒，皮肤接触会中毒，会损害器官。急救措施：吸入，将患者移到新鲜空气处，如呼吸停止进行人工呼吸，就医；皮肤接触，用肥皂和大量的水冲洗，立即将患者送往医院，就医；眼睛接触，谨慎起见用水冲洗眼睛；食入，禁止催吐，切勿给失去知觉者喂食任何东西，用水漱口，就医。储存注意事项：使容器保持密闭，储存在干燥通风处。打开了的容器必须仔细重新封口并保持竖放位置以防止泄漏。储存于室温。

2. 双氧水：外观为无色透明液体，是一种强氧化剂，其水溶液适用于医用伤口消毒、环境消毒和食品消毒。在一般情况下会分解成水和氧气，但分解速度极其慢。用于化学工业、纺织工业、造纸工业、医药和环境保护等，可用作氧化剂、漂白剂、消毒剂、脱氯剂；还可用于制造火箭燃料、有机或无机过氧化物、泡沫塑料等。急性毒性：可能引起燃烧或爆炸；强氧化剂。吞咽有害，造成严重皮肤灼伤和眼损伤，吸入有害。急救措施：吸入，迅速脱离现场至空气新鲜处，保持呼吸道通畅，如呼吸困难给输氧，呼吸、心跳停止立即进行心肺复苏术，就医；皮肤接触，立即脱去污染的衣着，用大量流动清水彻底冲洗，就医；眼睛接触，立即分开眼睑，用大量流动清水或生理盐水彻底冲洗 10～15min，就医；食入，用水漱口，禁止催吐，给饮牛奶或蛋清，就医。储存注意事项：与可燃物质、还原性物质、食品、饲料及强碱、金属分开存放。阴凉场所。保存在阴暗处。储存在通风的容器中。稳定后储存。

3. 甲醇：甲醇是无色有酒精气味易挥发的液体。主要用于制甲醛，在有机合成工业中用作甲基化剂和溶剂，是制造甲基叔丁基醚的原料，也可直接掺入汽油作为汽车燃料，还是制造某些农药、医药的原料。通常由一氧化碳与氢气反应制得。急性毒性：人口服中毒最低剂量约为 100mg/kg 体重，经口摄入 0.3～1g/kg 可致死。急救措施：吸入，新鲜空气，休息，给予医疗护理；皮肤接触，脱去污染的衣服，用大量水冲洗皮肤或淋浴，给予医疗护理；眼睛接触，先用大量水冲洗几分钟（如可能易行，摘除隐形眼镜），然后就医；食入，催吐（仅对清醒病人!），给予医疗护理。储存注意事项：耐火设备（条件）。与强氧化剂、食品和饲料分开存放。阴凉场所。

4. 乙酸乙酯：无色透明液体，低毒性，有甜味，浓度较高时有刺激性气味，易挥发，对空气敏感，能吸水分，使其缓慢水解而呈酸性反应。能与氯仿、乙醇、丙酮和乙醚混溶，溶于水（10%mL/mL）。能溶解某些金属盐类（如氯化锂、氯化钴、氯化锌、氯化铁等）反应。相对密度 0.902。熔点 −83℃。沸点 77℃。折射率 1.3719。闪点 7.2℃（开杯）。易燃。

蒸气能与空气形成爆炸性混合物。急性毒性：半数致死量（大鼠，经口）11.3mL/kg。急救措施：吸入，如果吸入，请将患者移到新鲜空气处，如呼吸停止进行人工呼吸，就医；皮肤接触，用肥皂和大量的水冲洗，就医；眼睛接触，用大量水彻底冲洗至少15min并就医；食入，禁止催吐，切勿给失去知觉者喂食任何东西，用水漱口，就医。储存注意事项：产品应储存于阴凉、通风的库房，仓温不宜超过30℃，防止阳光直接照射，保持容器的密闭。应与氧化剂、酸碱类等分开存放，储区应备有泄漏应急设备和合适的收容材料。

六、思考题

1. 推测的苯酚羟基化反应机理。
2. 苯二酚选择性计算方法是什么？
3. 提高苯酚转化率的方法是什么？
4. 列举其他制备苯二酚的方法并分析优缺点。
5. 说明双氧水氧化的优缺点。

Experiment 24　Preparation of hydroquinone from phenol catalyzed by titanium silicalite molecular sieve

1. Objectives

1.1　Understand the principle and application of TS-1 catalyst.

1.2　Understand the principle of TS-1 catalyst for phenol to hydroquinone.

1.3　Master the role of GC/MS in reaction monitoring and qualitative of product and by-product.

2. Principle

Phenol is mainly used to produce phenolic resin, caprolactam, and other important chemicals. Hydroquinone, including catechol and hydroquinone, is widely used in perfume, dye, and rubber synthesis. Up to now, 92% of phenol in the world is produced by the cumene process. However, the cumene process causes serious corrosion to equipment and environmental pollution. The production process of hydroquinone is mainly hydroxylation of phenol with hydrogen peroxide. Inorganic acid salt is used as a catalyst in the reaction. There are many problems such as difficult separation of catalyst and low conversion rate. It is of great theoretical and practical value to study the direct hydroxylation of phenol to hydroquinone under mild conditions.

In aqueous solution, hydrogen peroxide interacts with titanium active center of tetra coordination framework in titanium silicalite to form Ti-OOH active species. Then, phenol dissociated in the solvent attacks Ti-OOH and forms the intermediate of hydroquinone. Then the hydrogen atom on the benzene ring exchanges with the hydroxyl group in Ti-OOH to generate hydroquinone, which recovers the activity of the catalyst at the same time. According to the O-P mechanism, Mainly catechol and hydroquinone can be generated, and small amount of resorcin. With the extension of reaction time, hydroquinone may be further oxidized to benzoquinone (solution becomes darker). The reaction equation is as follows:

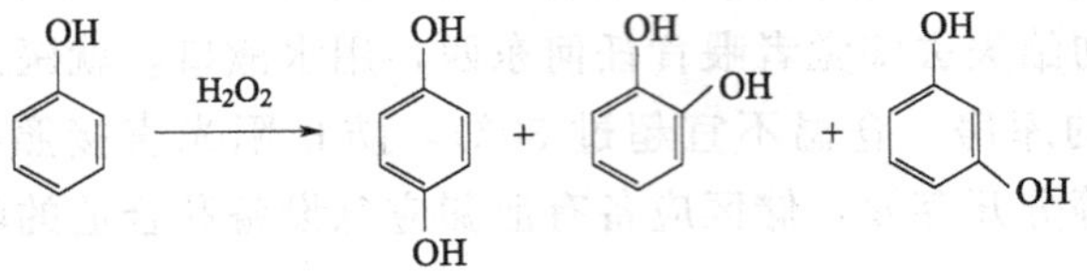

Fig. 1 The reaction equation of the experiment

3. Instruments and reagents

3.1 Instruments

Three port round bottom flask, magneton, electric heating sleeve, reflux condenser, thermometer, drip funnel, pear funnel, gas chromatograph (GC).

3.2 Reagents

Phenol, TS-1, hydrogen peroxide (30%), methanol, ethyl acetate.

4. Procedures

4.1 Experimental device

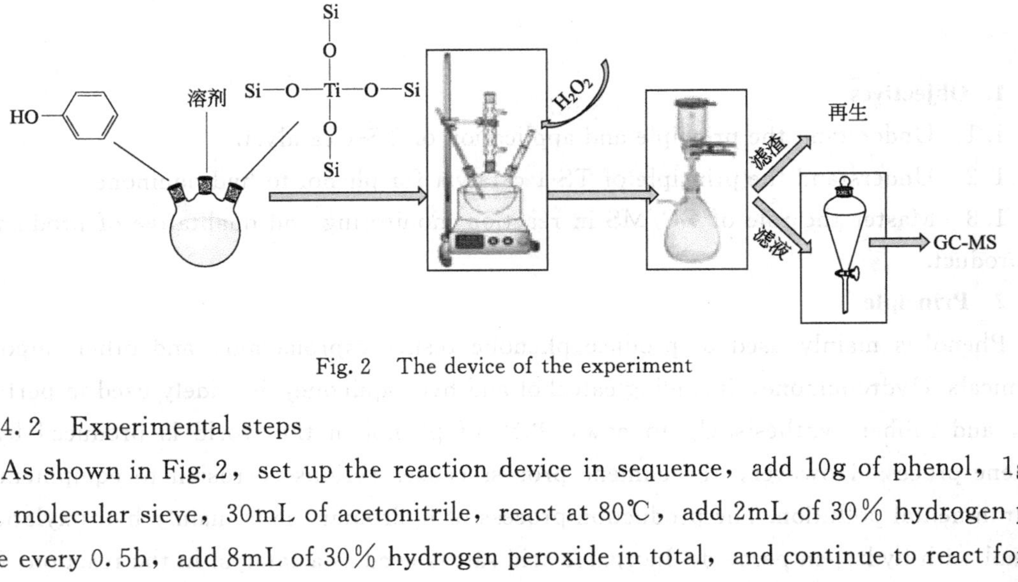

Fig. 2 The device of the experiment

4.2 Experimental steps

As shown in Fig. 2, set up the reaction device in sequence, add 10g of phenol, 1g of TS-1 molecular sieve, 30mL of acetonitrile, react at 80℃, add 2mL of 30% hydrogen peroxide every 0.5h, add 8mL of 30% hydrogen peroxide in total, and continue to react for 2h at 80℃. After the reaction, a small amount of methanol was added to quench the reaction, and the filtrate was filtered. The filtrate was analyzed qualitatively and quantitatively by GC-MS.

5. Experimental HSE

5.1 Phenol: phenol, also known as carbolic acid, hydroxybenzene, is the simplest phenolic organic matter, a weak acid. Normal temperature for a colorless crystal, toxic. Phenol is a common chemical and an important ingredient in the production of certain resins, fungicides, preservatives, and drugs such as aspirin. Phenol is corrosive, slightly soluble in water, and easily soluble in organic solution at room temperature. When the temperature is higher than 65℃, it can be soluble with water in any proportion. The solution is washed on the skin with alcohol. The phenol is pink when exposed to air. Acute toxicity: swallowing can be toxic. Inhalation can be toxic. Skin contact can be toxic. It can damage organs. First aid: inhalation: if inhaled, move the victim to fresh air. If breathing stops, give artificial respiration. Consult a doctor. Skin contact: rinse with soap and plenty of water. Take the patient to the hospital immediately. Consult a doctor. Eye contact: rinse eyes with water carefully. Ingestion: no vomiting. Do not feed anything to an unconscious person. Rinse your mouth with water. Consult a doctor. Storage precautions: keep container airtight and store in a dry and ventilated place. Open containers must be carefully re-sealed and held upright to prevent leakage. Store at room temperature.

5. 2 Hydrogen peroxide: it is a colorless transparent liquid and a strong oxidant. Its aqueous solution is suitable for the disinfection of medical wounds, environment, and food. It normally breaks down into water and oxygen, but it does so very slowly. Used for the chemical industry, textile industry, paper industry, medicine, and environmental protection, etc. , can be used as an oxidant, bleaching agent, disinfectant, dechlorination agent; It is also used to manufacture rocket fuel, organic or inorganic peroxides, foamed plastics, etc. Acute toxicity: may cause combustion or explosion; Strong oxidant. Swallowing is harmful. Caused severe skin burns and eye damage. Harmful by inhalation. First-aid measures: inhalation: quickly leave the scene to the fresh air. Keep your airway open. If breathing is difficult, give oxygen. Breathing, cardiac arrest, cardiopulmonary resuscitation immediately. Go to a doctor. Skin contact: immediately remove contaminated clothing and rinse thoroughly with plenty of running water. Go to a doctor. Eye contact: immediately separate eyelids and rinse thoroughly with plenty of flowing water or normal saline for 10～15min. Go to a doctor. Ingestion: gargle with water and do not induce vomiting. Drink milk or egg whites. Go to a doctor. Storage precautions: separate from combustible substances, reducing substances, food and feed, strong alkali, and metal. A cool place. Keep in a dark place. Store in a ventilated container. Storage after stabilization.

5. 3 Methanol: colorless liquid with a volatile odor of alcohol. Mainly used to make formaldehyde, in the organic synthesis industry as a methylating agent and solvent, is the manufacture of methyl tert-butyl ether raw materials, which can also be directly mixed into gasoline as car fuel, or the manufacture of some pesticides, medicine raw materials. It is usually made by reacting carbon monoxide with hydrogen. Acute toxicity: the lowest dose of oral intoxication is about 100mg/kg in human body weight. Oral intake of 0. 3～1g/kg can be fatal. First aid: inhalation: fresh air, rest, medical care. Skin contact: remove contaminated clothing, rinse skin or shower with plenty of water, and give medical care. Eye contact: Rinse with plenty of water for a few minutes (remove contact lenses if possible), then seek medical advice. Ingestion: induced vomiting (for conscious patients only!), giving medical care. Storage precautions: fire-resistant equipment (condition) . Store separately from strong oxidants, food, and feed. A cool place.

5.4 Ethyl acetate: colorless transparent liquid, low toxicity, sweet taste, high concentration with a pungent odor, volatile, sensitive to air, moisture absorption, slow hydrolysis, and acid reaction. Miscible with chloroform, ethanol, acetone, and ether, soluble in water (10%mL/mL). Can dissolve some metal salts (such as lithium chloride, cobalt chloride, zinc chloride, iron chloride, etc.) reaction. The relative density of 0.902. Melting point −83℃. The boiling point of 77℃. Refractive index 1.3719. Flashpoint 7.2℃ (open cup). Flammable. Steam can form explosive mixtures with air. Acute toxicity: half of the lethal dose (rats, orally) was 11.3mL/kg. First-aid: inhalation: if inhaled, move the victim to fresh air. If breathing stops, give artificial respiration. Consult a doctor. Skin contact: rinse with soap and plenty of water. Consult a doctor. Eye contact: rinse thoroughly with plenty of water for at least 15 minutes and consult a doctor. Ingestion: no vomiting. Do not feed anything to an unconscious person. Rinse your mouth with water. Consult a doctor. Precautions for storage: the product should be stored in a cool and ventilated warehouse, the temperature should not exceed 30℃, prevent direct sunlight, and keep the container airtight. It should be stored separately from oxidants, acids, and bases, and the storage area should be equipped with leakage emergency equipment and appropriate storage materials.

6. Questions

6.1 Speculated mechanism of phenol hydroxylation.

6.2 What is the calculation method of hydroquinone selectivity?

6.3 What is the methods to improve the conversion of phenol?

6.4 List other preparing methods of hydroquinone and analyze its advantages and disadvantages.

6.5 Please explain the advantages and disadvantages of hydrogen peroxide oxidation.

附　　录

附录一　X 射线衍射仪操作规程

应用 X 射线粉末衍射仪，对样品进行 XRD 测试。具体操作过程如下。

（1）开机前检查

检查仪器是否正常（循环水水位、管线开关、仪器电源等）。

（2）开机

依次打开循环水开关、主机背部空气开关和主机面板电源开关（绿色 POWER ON），注意查看（循环水水位、压力和温度设定值）。等待 10s，待控制面板左侧 Operate 灯亮并伴随稳定频率警告音时，按下 Doorlock 按钮。

（3）开射线及 X 射线光管老化（老化程序每天只需一次）

打开衍射仪控制 PC，用鼠标左键双击桌面上 MiniFlex Guidance 快捷图标，在弹出的登录界面中，于 Login name：处选择“Administrator”，并点击 OK 进入程序主界面。稍等 20s 左右，软件会与仪器获得正常通信并进行测角仪等的初始化。初始化结束后，在 Guidance 软件主界面左边，单击 Start up 按钮。在弹出的 Start up 窗口中，于 Generator usage 处选择 Use everyday，然后点击 Execute 按钮，系统会自动执行 XG 老化（Aging）。老化过程需要 7min30s，在此期间务必不能关闭 Guidance 软件！老化结束后，才能进入后面的样品测试步骤。

（4）样品测试

按 Doorlock 键，打开主机门，将样品片插入主机的样品槽中，关上主机门。在 Guidance 菜单栏中，找到第 3 项 Tasks→Package measurement。然后在 Guidance 软件主界面左上部，展开 General Measurement 折叠项，并点击 General Measurement，再单击红色方框中 1 General Measurement 黄色模块，即可打开测量主界面。

在测量主界面，设置文件名及保存路径（d：/data/导师姓名/学生名/文件名）并点击 Save 保存，继续输入测试条件并点击 OK 保存。

在确认完以上各项条件后，点击 OK 关闭所有输入的窗口，然后在 Guidance 软件主界面左边单击 Run 即可执行测试。

一个样品测试完毕后，按 Doorlock 键，打开主机门，取下第一个样品，重新装上第二个样品，按前述操作测试第二个样品，全部样品测试完毕将数据传到光盘。

（5）关机

样品测试结束后，在 Guidance 软件主界面左边，单击 Shut down 按钮，在设置窗口关闭 X 射线发生器，即可在 60s 左右关闭 X 射线（仪器顶部 X 射线指示灯熄灭），可以关闭软件。然后须等待 5min 左右，在仪器前面控制面板中部按下 Power off 白色按钮，然后在仪器背部左侧将电器开关打到 Off。最后关闭循环水机。

附录二　SEM 使用操作流程

一、检查和准备工作

1. 登记进入 SEM 测试间时间、测试样品种类和数量。

2. 确认屏幕右下角三个参数 Chamber pressure＜2.5×10^{-3} Pa，Gun pressure＜5×10^{-7} Pa，Emission current 约 60～100A（或屏幕右下角小瓶状图案全绿）即真空度正常。

3. 检查冷凝水温度正常（20℃±1℃）。

4. 检查温度湿度计（温度 20℃±5℃，湿度＜60%）。

5. 检查除湿机是否水满。

6. 凡是接触样品台、样品架及所有进入真空的零件，即制样、装样时，均需要戴手套操作（由于手上的油脂会污染镜筒且难以除去），放置样品后要用洗耳球吹一下确保样品不会被吹落；若样品台为自带样品座，需用少量乙醇擦拭干净后放入样品座内。严禁测试磁性样品！

粉末样品制样：擦净样品台，粘上导电胶，用牙签取少量样品涂于导电胶上，用洗耳球吹去浮着的粉末。

7. 装/换样前保证电子束（Beam on）处于关闭状态（灰色），点击 Vent 按钮，待 Vent 按钮由黄变灰后，轻轻拉出进样台（不用拉到底，拉出距离可以放置样品座即可），将样品座置于样品架上固定好，右手缓慢推入进样台（进样时注意观察屏幕右下窗口视频，防止样品撞上镜头），使进样台和机身轻轻合上，持续用右手向里推住进样台，同时左手点击 Pump 按钮，在观察到 Chamber pressure 数值开始下降后松开右手（此时可试着向外拉而不动）。

二、测试

1. 开 Beam：在样品室真空度到达 10^{-3} Pa 数量级时点击 Beam on（变黄）。

2. 升样品台：按住鼠标中键，点住样品台向上拖动至离探头大约 10mm 处停止，锁住 Z 轴。

3. 选电压/电流：根据样品选择所需要的电压和电流（金属材料可选择 20～30kV，喷金处理的高分子材料可选择 10～15kV，没有喷金处理的高分子材料可选择 10kV 以下；Spot 值一般选择 2.5 或 3）。在测试过程中，可更换电压，只是更换后需稳定一会儿，使电荷充分达到均匀分布。

4. 拍照：选择 ETD（二次探头，左上窗口）或者 CBS（散射电子探头，右上窗口）观察。

（1）倍数调整：点击相应窗口（鼠标左键点击）并取消暂停。在低倍数下找到样品，放大和缩小按钮分别为小键盘“＋”和“－”（或旋转操作台的 magnification 调倍数）；

（2）亮度/对比度调整：点击 F9 自动调节明暗和对比度（或旋转操作台的 Image Brightness/Contrast 调试）；

（3）焦距调整：再使用鼠标右键左右拖拽调焦（或旋转操作台的 Focus Corse/Fine 调试）；

（4）像散调整：当放大至×10000 倍以上时，若通过聚焦调试不清楚图像（或在过焦、欠焦时出现图像变形），可通过 shift＋鼠标右键左右拖拽调节像散（或旋转操作台的 Stigmator

X/Y 调试)，此时建议在小窗口下调节便于更快得到清晰的图像（F7 或点 Square 图标)；

(5) 保存图片：调节好以上各参数后，拍照可以选用快捷键 F2 慢扫或者 F4 快扫。

5. 更换样品座（针对小座)：先将 Z 轴解锁，然后双击需要去的 Hole X，接下来重复步骤 2。

6. 测试结束：先将放大倍数降低（30 左右)，关掉 Beam on（由黄变灰)，按住鼠标中键，点住样品台向下拖动至样品架底部（或将 *XYZ* 的数值改为零，点 Go To)。约 2min 后点击 Vent 按钮，待 Vent 按钮由黄变灰后，轻轻拉出进样台（不用拉到底，拉出距离可以取出样品座即可)，取出样品座后，右手缓慢推入进样台（进样时注意观察屏幕右下窗口视频，防止样品撞上镜头)，使进样台和机身轻轻合上，持续用右手向里推住进样台，同时左手点击 Pump 按钮，在观察到 Chamber pressure 数值开始下降后松开右手。

7. 拷取数据：按 Windows 键调出 Windows 界面，在外置刻录机上拷取数据到 DVD 盘或者 CD 盘（禁止使用 U 盘)，按照提示写入数据（Send to 光盘，将盘中等待写入的文件 Burn to disc，Next，Finish)。

附录三　Agilent 7890B/5977A 气质联用仪操作规程

一、开机

1. 打开载气（高纯氮气、氦气）钢瓶控制阀，设置分压阀压力至 0.5MPa（已设置好无须改变，只要打开钢瓶总阀即可)。

2. 依次打开 7890BGC、5975MSD 电源（若 MSD 真空腔内已无负压，则应在打开 MSD 电源的同时用手向右侧推真空腔的侧板直至侧面板被紧固地吸牢)，等待仪器自检完毕。

3. 打开计算机，桌面双击 GC-MS 图标，进入 MSD 化学工作站，软件自动连接仪器，下载操作参数、自动调谐文件等，进行初始化，等电脑界面提示初始化完毕。

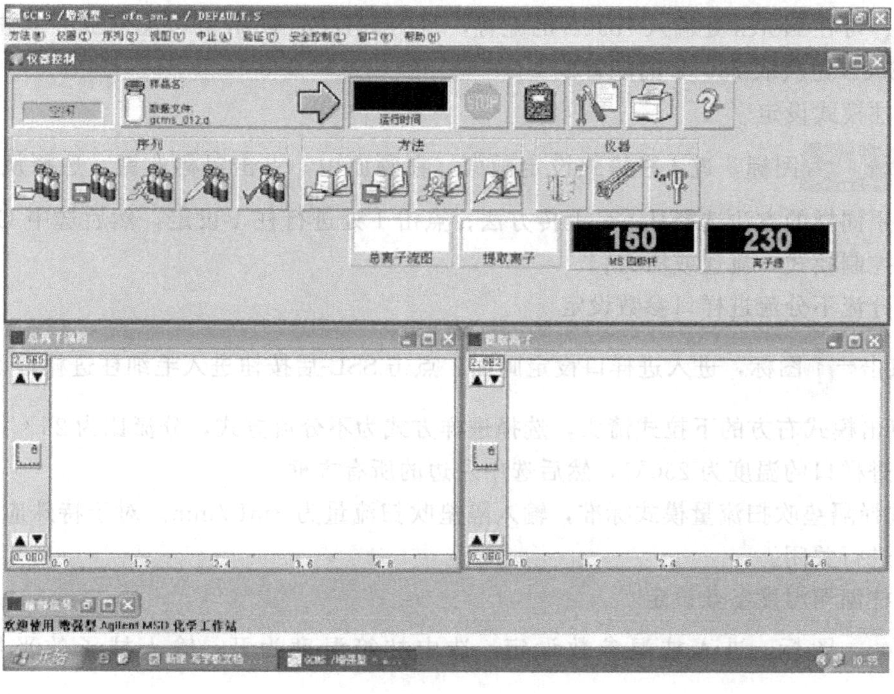

4. 在上图仪器控制界面下，单击视图菜单，选择调谐及真空控制进入调谐与真空控制界面，在真空菜单中选择真空状态，观察真空泵运行状态，此仪器真空泵配置为分子涡轮泵，状态显示涡轮泵转速，涡轮泵转速应很快达到100%，否则，说明系统有漏气，应检查侧板是否压正、放空阀是否拧紧、柱子是否接好。

二、调谐

调谐应在仪器至少开机2h后方可进行，若仪器长时间未开机，为得到好的调谐结果将时间延长至4h。

1. 首先确认打印机已连好并处于联机状态。

2. 在操作系统桌面双击GC-MS图标进入工作站系统。

3. 在上图仪器控制界面下，单击视图菜单，选择调谐及真空控制进入调谐与真空控制界面。

4. 单击调谐菜单，选择自动调谐MSD，进行自动调谐，调谐结果自动打印。

5. 如果要手动保存或另存调谐参数，将调谐文件保存到atune. u中。

6. 点击视图，然后选择仪器控制返回到仪器控制界面。

注意：

自动调谐文件名为atune. u。

标准谱图调谐文件名为stune. u。

其余调谐方式有各自的文件名。

三、样品测定

1. 方法建立

（1）7890B配置编辑

点击仪器菜单，选择编辑GC配置进入画面。在连接画面下，输入GC Name：GC7890B；可在Notes处输入7890B的配置，写7890B GC with 5977A MSD。点击获得GC配置按7890B钮获取7890B的配置。

（2）柱模式设定

点击图标，进入柱模式设定画面，在画面中，点击鼠标右键，选择从GC下载方法，再用同样的方法选择从GC上传方法；点击1处进行柱1设定，然后选中On左边方框；选择控制模式，流速或压力。

（3）分流不分流进样口参数设定

① 点击图标，进入进样口设定画面。点击SSL-后按钮进入毛细柱进样口设定画面。

② 点击模式右方的下拉式箭头，选择进样方式为不分流方式，分流比为25∶1，在空白框内输入进样口的温度为250℃，然后选中左边的所有方框。

③ 选择隔垫吹扫流量模式标准，输入隔垫吹扫流量为3mL/min。对于特殊应用亦可选择切换，进行关闭。

（4）柱温箱温度参数设定

点击图标，进入柱温参数设定。选中柱箱温度为开；输入柱子的平衡时间为

0.25min。

（5）数据采集方法编辑

从方法菜单中选择编辑完整方法项，选中除数据分析外的三项，点击确定。编辑关于该方法的注释，然后点击确定。

（6）编辑扫描方式质谱参数

① 点击图标编辑溶剂延迟时间以保护灯丝，调整倍增器电压模式（此仪器选用增益系数），选择要使用的数据采集模式，如全扫描、选择离子扫描等。

② 编辑 SIM 方式参数点击参数编辑选择离子参数，驻留时间和分辨率参数适用于组里的每一个离子。在驻留列中输入的时间是消耗在选择离子的采样时间。它的缺省值是 100ms。它适用于在一般毛细管 GC 峰中选择 2～3 个离子的情况。如果多于 3 个离子，使用短一点的时间（如 30ms 或 50ms）。加入所选离子后点击添加新组，编辑完 SIM 参数后关闭。

2. 采集数据

（1）点击 GC-MS 图标，在方法文件夹中选择所要的方法。

（2）选好方法后，点击

图标，依次输入文件名、操作者、样品名等相关信息，完成后按确定键，待仪器准备好后进样的同时按 GC 面板上的 Start 键，以完成数据的采集。

（3）当工作站询问是否取消溶剂延迟时，回答 NO 或不选择。如果回答 YES，则质谱开始采集，容易损坏灯丝。

3. 数据分析

（1）点击 GC-MS 数据分析图标，点击下图中文件调入数据文件。

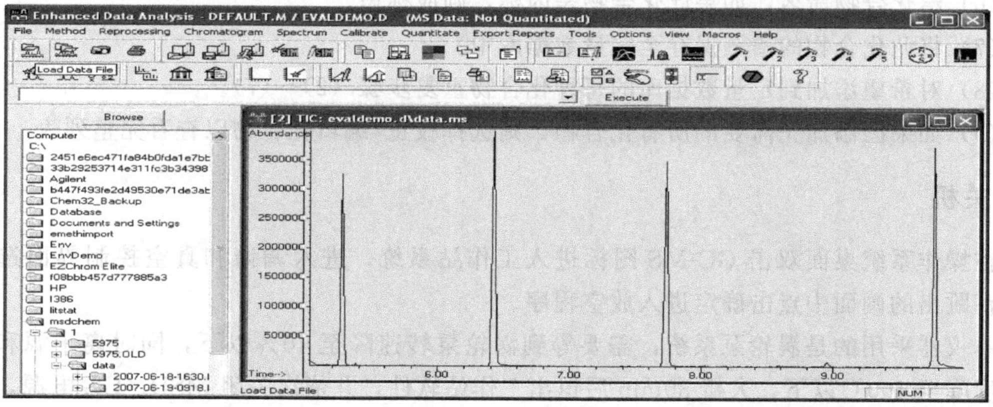

（2）在全扫描方法中要得到某化合物的名称，先右键双击此峰的峰高，然后再右键双击峰附近基线的位置得到本底的质谱图，然后在菜单文件下选择背景扣除即可得到扣除本底后该化合物的质谱图，最后右键双击该质谱图，便得到此化合物的名称。

（3）用鼠标右键在目标化合物 TIC 谱图区域内拖拽可得到该化合物在所选时间范围内的平均质谱图，右键双击则得到单点的质谱图。

（4）在选择离子扫描方法中不需要背景扣除操作。

4. 定量

定量是通过将来自未知量化合物的响应与已测定化合物的响应进行比较来进行的。手动设置定量数据库步骤如下所述。

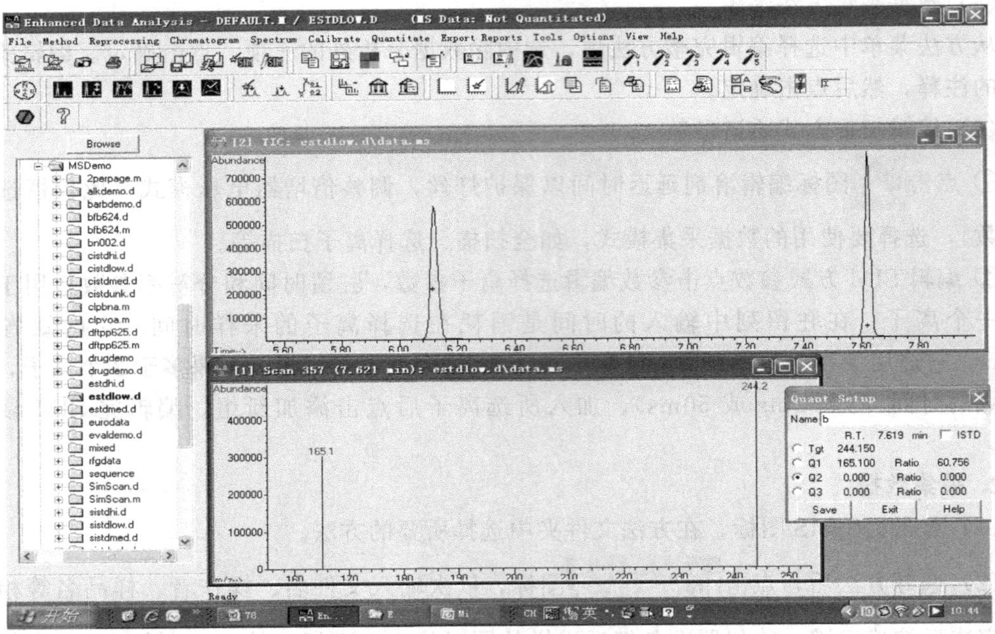

(1) 选择校正/设置定量访问定量数据库全局设置页。

(2) 手动检查由测定样品数据文件生成的色谱图。

(3) 通过单击色谱图中化合物的峰来分别选择每种化合物。

(4) 在显示的谱图中选择目标离子。

(5) 选择此化合物的限定离子。

(6) 给化合物命名，如果此化合物是内标，则应标识。

(7) 将此化合物的谱图保存至定量数据库中。

(8) 对希望添加到定量数据库的每种化合物重复步骤（2）～(7)。

(9) 如果已添加完需要的所有化合物，则选择校正/编辑化合物以查看完整列表。

四、关机

在操作系统桌面双击 GC-MS 图标进入工作站系统，进入调谐和真空控制界面选择放空，在跳出的画面中点击确定进入放空程序。

本仪器采用的是涡轮泵系统，需要等到涡轮泵转速降至 40%以下，同时离子源和四极杆温度降至 100℃以下，大概 30min 后退出工作站软件，并依次关闭 MSD、GC 电源，最后关掉载气。

附录四　高效液相色谱仪器操作规程

高效液相色谱是以高柱效小颗粒填料为载体，以高压泵为流动相驱动力，对物质进行快速分离和分析的方法，根据待分离组分与固定相和流动相相互作用强弱不同而实现时间和空间上的差异化分离，具有高压、高速、高效、高灵敏等特点。高效液相色谱法的应用远远广于气相色谱法，广泛用于合成化学、石油化学、生命科学、临床化学、药物研究、环境监测、食品检验及法学检验等领域。

一、高效液相色谱仪的组成

1. 系统结构简图（以福立 FL5090 为例）

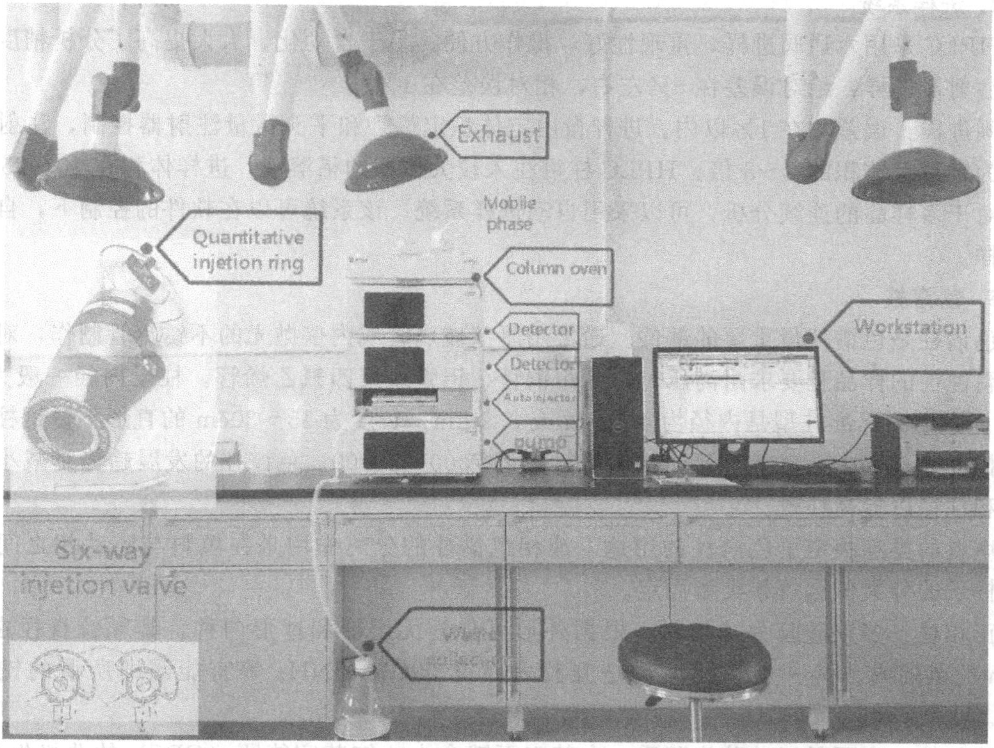

如上图所示，高效液相色谱主要由储液瓶、高压泵、进样器、色谱柱、检测器和色谱工作站组成。

2. 储液瓶和脱气机

主要用于存储流动相（溶剂或缓冲液）和除去流动相中的气泡。流动相脱气也通常预先采用超声方法脱气。

流动相进入高压泵之前必须脱气（脱气机一般包含在高压泵组件中），尤其是水和极性溶剂。否则气泡进入色谱柱以后，将影响分离效率、基线不稳使检测器灵敏度降低，甚至不能正常工作。

3. 高压泵

色谱柱用细粒填料，填充紧密，液体流动相黏度高，阻力大，必须高压输液。高压泵是高效液相色谱的关键部件，直接影响整个仪器和分析结果的可靠性。

高压泵分为两类：恒流泵和恒压泵。恒流泵：柱阻力变化，流动相黏度变化，引起柱压变化而流量恒定，如往复柱塞泵。恒压泵：流量可随外界阻力变化而输出压力恒定，例如装柱用的气动泵。HPLC 广泛采用往复柱塞泵，由液缸、柱塞、单向阀和驱动机组成。密封圈耐磨（PTF-石墨）单向阀，柱塞为人造红宝石。

泵流量不稳将影响化合物的保留时间，直接影响峰面积的重现性和定量精度，使噪音变大，最小检测量变大。双柱塞泵流量平稳，输液精度高，流量精度±0.1%，流量准确度±1%。双柱塞泵液缸容积很小（几微升至几百微升），适用于梯度洗脱。为提供无脉动的准确流量，在输液系统中还需加一些辅助设备：过滤器、混合器、阻尼器、测压装置。多元混合溶剂需用比例阀，如二元、三元和四元比例阀。溶剂按预定比例，低压混合后，经高压泵输入系统。

4. 进样系统

HPLC采用六通阀进样，重现性好，操作方便，易于自动化，大大提高了分析精度。

注射器进样：绝对误差在5%左右，相对误差在1%。

阀进样：误差均在1%以内。进样量由定体积定量管和平头微量注射器控制，注射器体积应为定量管体积的2～3倍。HPLC柱可注入较大体积的稀溶液，进样体积在10～250μL。

对于多样品的连续分析，可以采用自动进样系统；该系统可以在软件的控制下，自动逐个进样。

5. 色谱柱

色谱柱是色谱仪最重要的部件。通常用厚壁玻璃管或内壁抛光的不锈钢管制作，对于一些有腐蚀性的样品且要求耐高压时，可用铜管、铝管或聚四氟乙烯管。柱子内径一般为1～6mm。常用的标准柱型是内径为4.6mm或3.9mm，长度为15～30cm的直形不锈钢柱。填料颗粒度5～10μm，柱效以理论塔板数计大约7000～10000。色谱柱的发展趋势是减小填料粒度和柱径以提高柱效。

填料的类型决定了色谱柱的用途。液相色谱柱的分离作用是在填料与流动相之间进行的，柱子的分类依据填料类型而定。

正相柱：多以硅胶为柱填料。根据外形可分为无定形和球形两种，其颗粒直径在3～10μm的范围内。另一类正相填料是硅胶表面键合-CN、-NH_2等官能团即所谓的键合相硅胶。

反相柱：主要是以硅胶为基质，在其表面键合十八烷基官能团（ODS）的非极性填料。也有无定形和球形之分。

常用的其他的反相填料还有键合C8、C4、C2、苯基等，其颗粒粒径在3～10μm之间。

6. 检测器

高效液相色谱仪的检测器主要用于分析样品信号的采集，常用的检测器有紫外检测器、荧光检测器、示差检测器、电导检测器、光电二极管阵列检测器等。

（1）紫外检测器

紫外检测器利用被分析样品对特定波长紫外光的选择性吸收，组分浓度与吸光度的关系遵守比尔定律。紫外检测器是最常用的检测器，应用最广，对大部分有机化合物有响应，具有灵敏度高、线性范围宽等优点。

（2）荧光检测器

荧光检测器是一种高灵敏度、高选择性检测器。对多环芳烃、卟啉类化合物、维生素B、黄曲霉素、农药、药物、氨基酸、甾类化合物等有响应。其结构及工作原理与荧光光度计相似。

（3）示差检测器

示差检测器是除紫外检测器之外应用最多的检测器。示差折光检测器是通过连续测定流通池中溶液折射率的方法来测定试样浓度的检测器。溶有样品的流动相和纯流动相之间折射率之差表示试样在流动相中的浓度。

（4）电导检测器

其作用原理是根据物质在某些介质中电离后所产生的电导变化来测定电离物质含量。

（5）光电二极管阵列检测器

阵列由1024个光电二极管组成，每个光电二极管宽仅50μm，各检测一窄段波长。在检测器中，光源发出的紫外或可见光通过液相色谱流通池，在此流动相中的各个组分进行特征吸收，然后通过狭缝，进入单色器进行分光，最后由光电二极管阵列检测，得到各个组分的吸收信号。经计算机快速处理，得三维立体谱图。

7. 色谱工作站

色谱工作站主要完成色谱信号的在线采集、检测、调整和数据处理，不同厂家的高效液相色谱配备有不同的色谱工作站。

二、实验安全要求与注意事项

1. 安装仪器的房间应通风良好，安装有排气装置，否则会引起中毒或不良刺激，也可能引起火灾，因为液相色谱仪所使用的溶剂大多是易燃且有毒的。

2. 严禁在仪器附近吸烟，使用明火或其他火源，或安装其他任何发射或可能发射火花的设备，房间内应配备灭火装置或设备。

3. 房间内应配备自来水龙头或其他冲洗设备。如果溶剂进入眼睛或有毒溶剂溅到皮肤上，应立即用清洁的水冲洗。

4. 避免在有腐蚀性气体或大量粉尘的地方安装本仪器，否则会影响仪器正常运转，并缩短仪器使用寿命。

5. 实验室温度需控制在20℃左右，建议安装空调；电源电压为交流220V，功率为200W；房间内有良好的接地线。

三、操作规程

1. 准备

（1）使用前应根据待检样品的检验方法准备所需的流动相，流动相必须使用0.45μm滤膜过滤，并用超声波清洗仪脱气10～15min；准备好样品和标准溶液（也可在平衡系统时配制）；更换合适的色谱柱（注意色谱柱进出口位置应与流动相流向一致，不要装反）。

（2）通电前应检查仪器设备之间的电源线、数据线和输液管道是否连接正常。

2. 开机和仪器操作

（1）开机

插上仪器的电源插头，打开仪器后面板上的电源开关，仪器进行自检。打开打印机、电脑显示器和主机。打开色谱工作站，选择正确的数据通道。

（2）设置仪器参数

按照样品的检验方法，依次设置好检测器的波长、响应时间、量程和泵的流量，并设置检测器的基线，一般设为5000mV左右。

（3）排气泡

打开排空阀，设置泵的流量为5mL/min（注意：必须确认排空阀是打开的，否则可能损坏安装好的色谱柱），启动泵，冲洗2～5min（或更长时间）后停泵，将泵的流量改回原来的设置，关闭排空阀，排气泡操作完成。

3. 平衡系统

（1）检查管路连接、柱接口及泵头处是否漏液，如漏液应予以排除。

（2）换检验方法规定的流动相冲洗系统，观察基线和压力变化。如果冲洗至基线和压力平稳时，可认为系统已达到平衡状态，可以进样。

4. 进样

（1）用针头过滤器过滤试样溶液（注射液可不必过滤），用试样溶液清洗进样注射器，排除气泡后抽取适量。

（2）用定量环定容进样时，进样量应不小于定量环体积的 2 倍（5～10 倍准确性更佳）。

（3）将进样阀手柄转动至 Load 位置，将进样注射器针完全插入进样阀入口中，平稳地注入试样溶液，除另有规定外，让进样注射器留在进样阀上，将进样阀手柄快速转动至 Inject 位置，系统将自动运行，采集数据并记录图谱。

（4）让进样阀手柄保持在 Inject 位置，到下次进样前 1～2min 切换回 Load 位置，将注射器从进样阀中拔出。

（5）先用水再用试样溶液清洗进样注射器后，按以上程序继续进样，直至完成。

5. 清洗进样阀

（1）用注射器吸 10mL 超纯水，将注射针导入口冲洗头连接到注射器出口上（不要针），并将它们一起接到进样口上。

（2）使进样阀保持在 Inject 位置，慢慢将水推入，水将通过注射针导入口、引导管、注射针导入管和注射针密封圈，由样品溢出管排出。

（3）将进样阀换成 Load 位置，同样清洗一次。

6. 清洗柱

（1）C_{18}柱先用超纯水∶甲醇（95∶5）以 1mL/min 冲洗 60min 以上，再用甲醇或乙腈冲洗 30min 左右。

（2）用水冲洗色谱柱后，再用 60mL 超纯水冲洗柱塞杆。

（3）清洗完成后，关闭仪器的电源开关，断开仪器电源。

7. 填写仪器使用记录

附录五　电极的使用说明

一、电极的用途

本电极是玻璃电极和参比电极组合在一起的塑壳可充式复合电极，是 pH 测量元件，用于测量水溶液中氢离子活度（pH 值），它广泛用于化工、轻工业、医药工业、染料工业和科研事业中的酸碱度检测。

二、电极的使用维护及注意事项

1. 电极在测量前必须用已知 pH 的标准缓冲溶液进行电位校准，为了取得正确的结果，已知 pH 要可靠，而且 pH 越接近，被测值越好。

2. 取下帽后要注意，在塑料保护栅内的敏感玻璃泡不宜与硬物接触，任何破损和擦毛都会使电极失效。

3. 测量完毕，不用时应将电极保护帽套上，帽内应放少量补充液，以保持电极泡的湿润。

4. 复合电极的外参比补充液为 3mol/L 氯化钾溶液（附件有：内装 3mol/L 氯化钾小瓶一只，用户只需加入 20mL 蒸馏水摇匀，此溶液即为外参比溶液）。补充液可以从上端小孔加入。

5. 电极的引出端，必须保持清洁和干燥，绝对防止输出两端短路，否则将导致测量结果失效或失准。

6. 电极应与输入阻抗较高的酸度剂（$\geqslant 10^{12}\Omega$）配套，能使电极保持良好的特性。

7. 电极避免长期浸在蒸馏水中或蛋白质溶液和酸性氟化物溶液中，并防止与有机硅油脂接触。

8. 电极经长期使用后，如发现梯度略有下降，则可把电极下端浸泡在 4% HF 中 3～5s。用蒸馏水洗净，然后在氯化钾溶液中浸泡，使之复新。

9. 被测溶液中如含有易污染敏感球泡或堵塞液接界的物质，而使电极钝化，其现象是敏感梯度降低，或读数不准。此时，则应根据污染物的性质，以适当溶液清洗，使之复新。

注：选用清洗剂时，如用能溶解聚碳酸树脂的清洗液（如用四氯化碳、三氯乙烯、四氢呋喃等），则可能把聚碳酸树脂溶解后，附着在敏感玻璃泡上，而使电极失效，请慎用。

附录六　奥立龙 CHN060 型酸度计的操作规程

1. 双缓冲溶液标定

每次接通电源后需对主机预热 10min，然后对电极进行标定 1 次，保证仪器处于最佳工作状态。

（1）根据被测溶液的酸碱性，选 2 点缓冲溶液值进行标定。第一点缓冲溶液的 pH 应为 6.86，第二点缓冲溶液的 pH 需与被测溶液的酸碱性一致，即选择 pH＝4.00 或 pH＝9.18 的缓冲溶液。

（2）选区按标定键（CAL），然后按确定键（YES）。通过滚动键选择二点缓冲溶液（4～7 或 7～9），再按确定键（YES）。

（3）蒸馏水冲洗电极，并用滤纸擦干，然后将电极置于 pH＝6.86 的缓冲溶液，搅动一下缓冲溶液，当 Ready 灯亮时，按确定键（YES）表示第一点已标定。

（4）蒸馏水冲洗电极，并用滤纸擦干，然后将电极置于 pH＝4.00 或 pH＝9.18 的第二点缓冲溶液，搅动一下缓冲溶液，当 Ready 灯亮时，按确定键（YES）表示第二点已标定。

（5）两点标定结束后，仪器进入斜率方式，并显示斜率值。

2. pH 的测量

（1）标定结束后，用蒸馏水冲洗电极。

（2）将电极放入被测溶液，并轻微搅动一下，当 Ready 灯亮时，记录 pH。

3. 储藏电极

储藏一周以内，应将电极浸泡于 3mol/L KCl 溶液中，避免溶液挥发。切勿将电极储藏于蒸馏水中，否则会缩短电极寿命。长期储藏时，应用蒸馏水冲洗电极，需用保护盖将探测表面盖好并存放于干燥处。

附录七　显微熔点仪操作规程

1. 样品的制备

（1）对待测的样品进行干燥处理；样品烘干时，温度应控制在待测物品的熔点温度以下。

（2）将干燥后的待测样品研细，备用。

2. 开机部分

（1）检查连接：a. 热台的电源线是否接入调压测温仪的输出端，b. 传感器是否插入热台孔，c. 传感器是否与调压测温仪连接。

（2）将调压测温仪的电源线插入插座，接入电源。

（3）将温控旋钮 1 和 2 向左归零，打开电源开关。

3. 测样操作

（1）取适量待测样品（不大于 0.1mg）放在一片载玻片上，并使药品分布薄而均匀，盖上另一片盖玻片，轻轻压实，然后放在热台中心。

（2）盖上隔热玻璃，保护显微镜镜头。

（3）上下调整显微镜，直至能清晰地看到待测物品的像。

（4）根据样品熔点的温度值，控制调温钮 1 和 2（1-升温速率快，2-升温速率慢），在测试过程中，前段温度升温迅速，中段升温渐慢，后段升温平缓（当温度接近样品熔点温度 40℃左右时，使升温速率减慢。当接近样品熔点温度 10℃左右时，调整温控旋钮使升温速率约 1℃/min）。

（5）观察样品的融化过程，记录初融和全融时的温度值，即完成一次测试；

（6）多次测试，计算平均值。

4. 关机部分

（1）测试完毕后，用镊子取下隔热玻璃和盖玻片，放在隔热垫上，将温控旋钮 1 和 2 向左归零，关闭电源开关；测试过程中，一定要使用镊子夹持放入或取出样品，热台属高温部件，严谨用手触摸，以免烫伤。

（2）待热台冷却后，及时切断电源，将仪器清理复原，用过的盖玻片放入酒精中浸泡后擦拭干净，以备下次使用。

附录八　CHI 660 电化学工作站操作规程

1. 打开电脑，打开电化学工作站。
2. 电路连接，绿色铁夹接工作电极，红色铁夹接辅助电极，白色铁夹接参比电极。
3. 双击打开电脑桌面上的软件（CHI 660），打开软件界面。
4. 点击 Setup，选择 Technique，选择 CV-Cyclic Voltammetry。

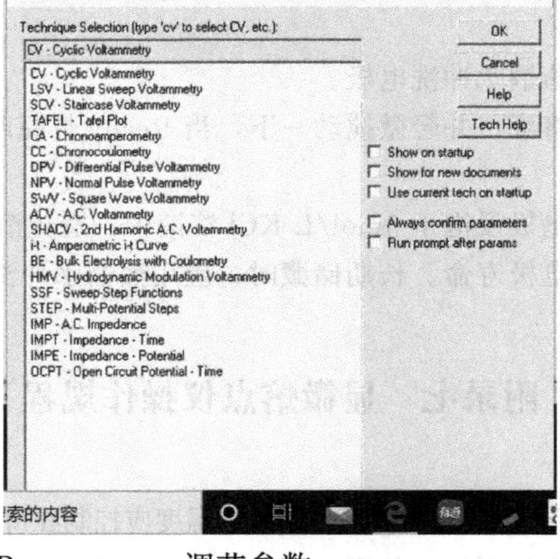

5. 点击 Setup，选择 Parameters，调节参数。

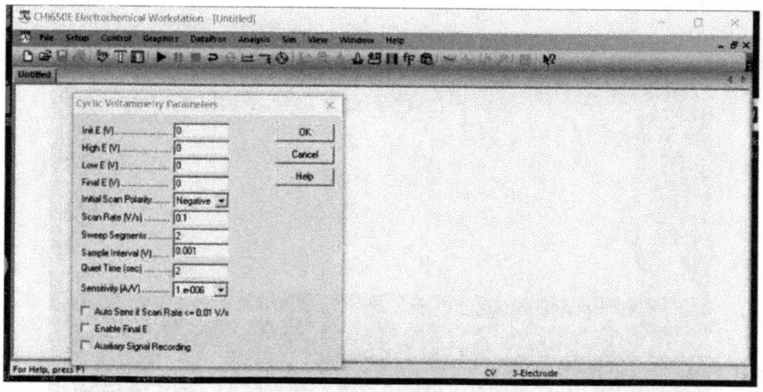

Init E　　　　−0.2V
High E　　　　0.9V
Low E　　　　−0.2V
Final E　　　　0V

如果出现电流溢出（OVER FLOW），将灵敏度（Sensitivity）调高。

6. 点击 Control，选择 Run experiment，界面右上角显示剩余时间。

7. 实验结束，剩余时间消失，将实验结果另存为目标文件，此文件类型为工作站默认类型，Excel 无法打开。

8. 打开目标文件下的实验图形，打开数据处理下的“查看曲线”，选择显示曲线（不选第一次循环），确定。出现列表对话框，点击保存，保存类型为 Excel。

附录九　Bruker 400MHz 核磁共振谱仪操作规程

核磁共振波谱是一种功能强大、非破坏性的结构分析方法，主要用于确定分子结构和研究反应动力学。可用于有机合成产物及其中间体的成分和结构测定；天然产物结构测定；医药、农药、兽药结构确证；高分子产品和原料的测定；有机磷化合物、含磷农药的 ^{31}P 核磁图谱测定；混合物中有机成分的分析，包括定性和相对近似含量；此外，用同核（^{1}H-^{1}H）及异核（^{1}H-^{13}C/^{15}N）二维等多维核磁共振方法可测定有机小分子的空间结构。

一、Bruker 400MHz 核磁共振波谱仪的组成

1. 磁体

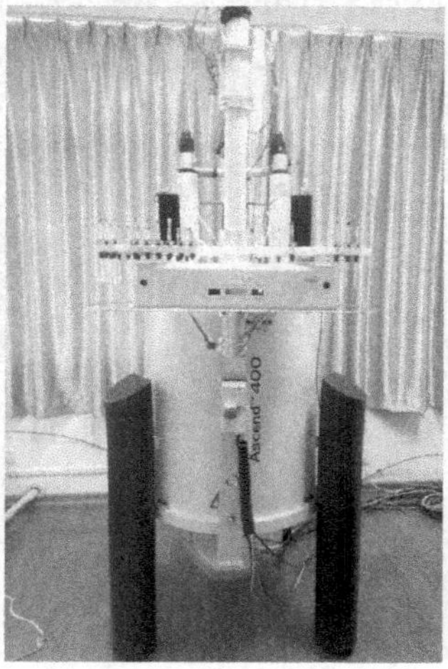

2. 谱仪

3. 电脑

4. 空气压缩机

5. 自动进样器、核磁管、转子及转盘

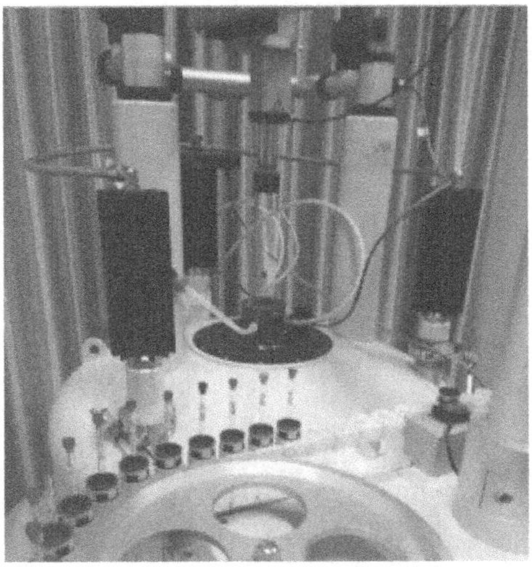

6. 液氮罐

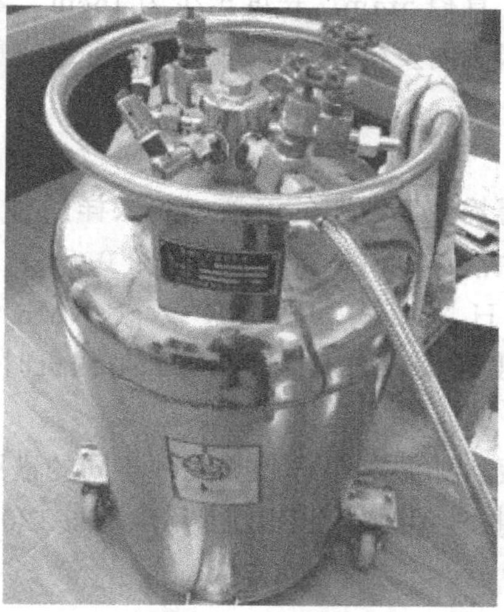

二、实验安全要求与注意事项

严禁携带电子、金属设备（例如心脏起搏器、人工关节、手表、钥匙等）进入核磁共振实验室，否则所携带的设备会发生故障。

三、操作步骤

1. 样品准备

（1）样品纯度

进行核磁测试的样品纯度一般应≥95%，无铁屑、灰尘、滤纸毛等杂质。

（2）氘代溶剂

本仪器配置仅适合做液体样品。要求样品在某种氘代溶剂中有良好的溶解性能。常用的氘代溶剂：$CDCl_3$、D_2O、d^6-DMSO、d^6-Acetone、C_6D_6、CD_3CN、CD_3OD、d^8-THF 等。

（3）优质核磁管

样品测试需自备样品管，直径 5mm，长度至少为 16cm，要求管内外壁干净，管壁无划痕破损（严禁样品管在仪器探头内发生断裂，一旦断裂将造成重大仪器故障，需支付高昂的维修费用）。

（4）样品溶解

将样品与溶剂溶解，制得均一、透明的溶液，装入样品管中，如有不溶物可超声助溶，并过滤后装入，悬浊液和乳浊液不能测试。溶剂用量：0.5～0.6mL（高度 4cm）（溶剂太少不能正常匀场）。样品用量：^{1}H NMR、^{19}F NMR 及^{31}P NMR 谱需样 10～20mg；^{13}C NMR 谱 30～50mg（因^{13}C 天然丰度较小，增加样品量可以节约机时和测试费用）。

（5）样品编号

不同的样品编号不能重复（如果一个样品测几种不同的谱图则用同一个编号，请在送样单里标注所要测试的内容）。推荐编号：lisi-12、ls-15、lisi-120501、ls-120501-4……

（6）数据的处理

用户需要自行处理谱图。

2. 开机和仪器操作

（1）打开空气压缩机的气流。

（2）用量深量筒确定插在核磁管上转子的位置，待用。

（3）命令栏中输入“ej”后按回车，听到气流声后，将带转子的核磁管放入磁体顶端。

（4）命令栏中输入“ij”后按回车，将核磁管送入磁体，气流声消失。

（5）命令栏中输入“edc”后按回车，对话框里输入样品名和实验名称，点击“OK”。

（6）锁场：命令栏里输入“ock”后按回车，选择所用的氘代溶剂后点击“OK”；直至命令栏下方出现“ockn：finished”方可进行下一步操作。

（7）调谐：命令栏里输入“atma”后按回车，直至任务栏出现“Job Succeed (Command ‘atma’ on)……”成功后方可进行下一步操作。

（8）匀场：命令栏里输入“topshim”后按回车，直至任务栏出现“Job Succeed (Command ‘topshim’ on)……”成功后方可进行下一步操作。

（9）扫描采样。

① 扫描^{1}H 谱，命令栏内输入“ns 8；getprosol；rga；zg；1H”后按回车；

② 扫描^{19}F 或者^{31}P 谱，命令栏内输入“getprosol；rga；zg；1H”后按回车；

③ 扫描^{13}C 谱，命令栏内输入“ns 1024；getprosol；rga；zg”后按回车；确定已扫描次数可在命令栏里输入 13C 回车，然后再一次在命令栏里输入 13C 回车，若谱图质量达标，最后在命令栏里输入“stop”停止采样。

（10）待扫描采样结束后，命令栏内输入“ej”按回车，将核磁管弹出后，轻轻取出，将转子放回原处，并在命令栏内输入“ij”按回车，关闭腔体顶端气流。

3. 填写仪器使用记录

（1）结束测试，认真填写使用记录表。

（2）若接下来无人继续测试，关闭空气压缩机电源。

附录十 FluoroMax-4 荧光光谱仪操作规程

荧光光谱仪，又称荧光分光光度计，是一种定性、定量分析的仪器。通过荧光光谱的测试可以获得物质的激发光谱、发射光谱、荧光寿命以及液体样品浓度等方面的信息。荧光光谱仪的组成在硬件上是相似的，其差别主要在于硬件材质、仪器精密程度以及部分次要功能上的不同。荧光光谱仪的重要的部件包括光源、激发单色器、发射单色器、光电倍增光（Photomultiplier，PMT）以及外联设备电脑等。图 1 为荧光光谱仪的主要组成示意图。通过荧光光谱仪的检测，可以获得物质的激发光谱、发射光谱、量子产率、荧光强度、荧光寿命、斯托克斯位移、荧光偏振与去偏振特性以及荧光的猝灭方面的信息。

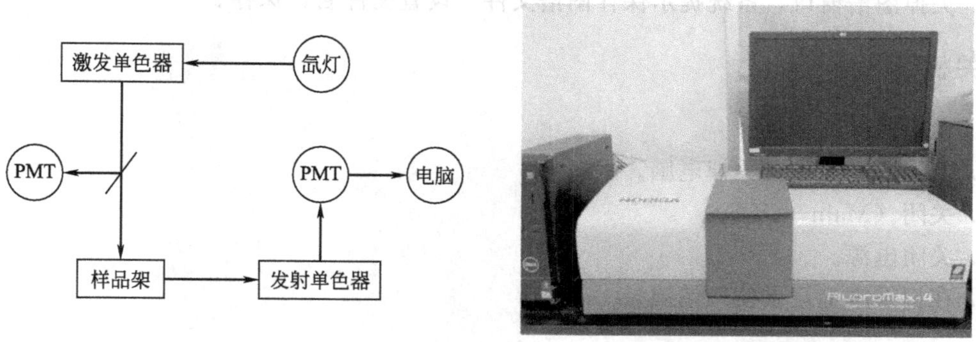

图 1　荧光光谱仪的组成

操作规程如下。

一、开机前准备

1. 实验室温度应保持在 15～30℃之间，湿度应保持在 45%～70%之间。
2. 确认样品室内无样品后，关上样品室盖。

二、开机（开机，关机顺序千万不能搞错）

1. 打开电源开关（POWER→ON）待风扇正常运转。
2. 打开（X LAMR START）按钮，黄色指示灯亮（注意按钮时间不能超过 20s）。
3. 在氙灯点亮后，等待 5～10min，打开（MAIN→ON）。
4. 开计算机及外设，点击桌面上的 Flsolutionc 软件，等待光谱仪自检。

三、编辑分析方法

1. 点击右边菜单的 Methed 按钮。
2. 出现一个 Analysis methed 对话框。
3. 点击 General 按钮。
4. 在 Measurement 中可选择（波长扫描 Wavelength scan；时间扫描 Time scan；光强度 Photometry；三维扫描 3D scan）。
5. 点击（Instrument），可选择荧光激发 Excitation、荧光发射 Emission、同步 Synchronous。可选择荧光 Fluorescence，发射光 Luminescence、磷光 Phosphorescence，可选择扫描波长、扫描波长范围、扫描速度、电压、狭缝。
6. 点击（Quantitation）可选择波长、峰面积、峰高、导数、比率。
7. 分析方法建立后，点击 OK，方法生成。

四、测试样品

1. 样品放入样品室后，点击 Measure 按钮，开始测量。
2. 检测完成后。

五、打印报告，保存文件

1. 点击 Report 按钮，打印报告。
2. 关掉图谱窗口，系统提示保存图谱文件，设置文件名，保存。

六、关机

1. 拿出样品池。
2. 退出所有窗口，关掉电脑。
3. 关闭（Main→OFF）。
4. 关闭电源。
5. 在记录本上记录使用情况。

参 考 文 献

[1] 魏星跃，方玲，李宁．应用化学专业实验［M］．成都：西南交通大学出版社，2016.

[2] 田维亮．化学工程与工艺专业实验［M］．上海：华东理工大学出版社，2015.

[3] 刘玉婷，梁钢涛，尹大伟．2-氨基-1,3,4-噻二唑合成新工艺及优化［J］．陕西科技大学学报，2014，32（2）：60-64.

[4] 宋健，王军波，鹿明，等．钛硅分子筛应用于苯二酚合成的研究［J］．化学工业与工程，2002，19（2）：159-162.

[5] 王彦红，王德松，李雪艳，等. 溶胶-凝胶法制备纳米 TiO_2 及其光催化活性的研究［J］．河北科技大学学报，2007，28（3）：202-205.

[6] 慕春红，刘鹏，朱刚强，等. ZnS：Cu 纳米微粒的光致发光特性［J]. 无机化学学报，23（5）：844-849.

[7] 陈虹，李莉．化工专业实验［M］．2 版．重庆：重庆大学出版社，2013.

[8] 朱红军，刘睿．有机化学微型实验［M］．3 版．北京：化学工业出版社，2018.

[9] 朱莉娜，孙晓志，弓保津．高校实验室安全基础［M］．天津：天津大学出版社，2014.

[10] 钱人元，等．黏度法测高聚物分子量［J］．化学通报，1955，7：396.

[11] 复旦大学高分子教研组．高聚物的分子量测定［M］．上海：上海市科学技术编译馆，1965.

[12] 虞志光．高聚物分子量及其分布的测定［M］．上海：上海科学技术出版社，1984.